FLAMMABILITY AND SENSITIVITY OF MATERIALS IN OXYGEN-ENRICHED ATMOSPHERES: SECOND VOLUME

A symposium sponsored by
ASTM Committee G-4 on
Compatibility and Sensitivity
of Materials in Oxygen-Enriched Atmospheres
Washington, DC, 23–24 April 1985

ASTM SPECIAL TECHNICAL PUBLICATION 910
Michael A. Benning, Air Products and
Chemicals, Inc., editor

ASTM Publication Code Number (PCN)
04-910000-17

 1916 Race Street, Philadelphia, PA 19103

Library of Congress Cataloging-in-Publication Data
(Revised for vol. 2)

Flammability and sensitivity of materials in oxygen-enriched atmospheres.

(ASTM special technical publication; 812)
Vol. 2: Symposium held in Washington, D.C., April 23–23, 1985; edited by Michael A. Benning.
Includes bibliographical references and index.
1. Materials—Flammability—Congresses. 2. Oxygen
index of materials—Congresses. I. Werley, B. L. (Barry
(Barry L.) II. Benning, Michael A. III. ASTM Committee
G-4 on Compatibility and Sensitivity of Materials in
Oxygen-Enriched Atmospheres. IV. Series: ASTM special
technical publication; 812, etc.
TH9446.3.F56   1983   628.9′22   82-73766
ISBN 0-8031-0474-X (v. 2)

NOTE

The Society if not responsible, as a body,
for the statements and opinions
advanced in this publication.

Printed in Baltimore, MD
May 1986

Foreword

The symposium on Flammability and Sensitivity of Materials in Oxygen-Enriched Atmospheres: Second Volume was presented at Washington, DC, 23–24 April 1985. The symposium was sponsored by ASTM Committee G-4 on Compatibility and Sensitivity of Materials in Oxygen Enriched Atmospheres. Michael A. Benning, Air Products and Chemicals, Inc., and John O. Cronk, AIRCO, Inc., served as chairmen of the symposium. Michael A. Benning is also editor of the resulting publication.

Related ASTM Publications

Fire Safety: Science and Engineering, STP 882 (1985), 04-882000-31

Behavior of Polymeric Materials in Fire, STP 816 (1983), 04-816000-31

Flammability and Sensitivity of Materials in Oxygen-Enriched Atmospheres, STP 812 (1983), 04-812000-17

Fire Risk Assessment, STP 762 (1982), 04-762000-31

Fire Standards and Safety, STP 614 (1977), 04-614000-31

Ignition, Heat Release, and Noncombustibility of Materials, STP 502 (1972) 04-502000-31

A Note of Appreciation
to Reviewers

The quality of the papers that appear in this publication reflects not only the obvious efforts of the authors but also the unheralded, though essential, work of the reviewers. On behalf of ASTM we acknowledge with appreciation their dedication to high professional standards and their sacrifice of time and effort.

ASTM Committee on Publications

ASTM Editorial Staff

Susan L. Gebremedhin
Janet R. Schroeder
Kathleen A. Greene
William T. Benzing

Contents

Overview

The purpose of this second symposium on the flammability and sensitivity of materials in oxygen atmospheres was to build on the base set by the first symposium, documented in STP 812. As with the first symposium, the aim was to generate a collection of papers that would

- provide a reference text on a subject that is not widely addressed in accessible literature,
- build a reference for both the skilled and the uninitiated in the concepts and practices for the design of oxygen systems,
- provide a data base to support the use of ASTM Committee G-4 guides and standards.
- serve as a guide to Committee G-4 in their future efforts to address the problems of safety in the use of oxygen.

These papers, together with those from the first symposium, show clearly that much work remains to be done to understand many aspects of materials selection, system design and maintenance for the maximum safety in handling gases rich in oxygen. However, they do provide an excellent foundation that documents the most current philosophies in these areas among practicing professionals, and they contain much useful data not otherwise readily available. These sources, for both the experienced and the novice, should continue to grow as more symposia are held, and they will provide a guide for future work.

The papers presented covered three key areas of importance in the design and operation of oxygen systems. Six of the fourteen papers address various aspects of the ignition of materials, both metallic and nonmetallic. These include reports on tests to simulate postulated mechanisms for ignition, reports on tests to measure ignition properties, and a theoretical analysis of the ignition process. Three of the papers cover the flammability of metals in pure and dilute oxygen. Five papers cover the related topics of cleanliness and contamination. In addition, the keynote address provides an excellent review of current practice for the selection of materials for oxygen service, including references to both American and European standards.

Ignition of Materials

Two of the papers relating to the ignition of materials report on test programs by the National Aeronautics and Space Administration (NASA) to investigate what are thought to be common ignition methods for metals in oxygen systems. The two methods are particle impact and frictional heating. The purpose of the tests was to rank metals according to their resistance to each type of ignition. Another paper reports on the ignition of metals by laser heating to investigate the ignition and combustion properties of metals and to gain a better understanding of these processes. A further paper reports a theoretical analysis of the ignition process for metals and attempts to apply the developed model to the data for ignition by frictional heating already noted. The two remaining papers report the use of new, instrumental techniques to determine the ignition properties of nonmetals.

Ignition of Metals and Alloys

Benz, Williams, and Armstrong reported on ignition of metals by high velocity particles in tests made at NASA White Sands Testing Facility (WSTF). In the tests, metal targets were impacted by aluminum or Type 304 stainless steel, spherical particles in oxygen at elevated temperatures, and pressures. Aluminum, Types 304 and 316 stainless steels and Inconel 718 were ignited by aluminum particles, but Monel 400 was not. More extreme conditions were required for ignition by stainless steel particles than by aluminum particles of the same size. The results indicated that ignition and burning of the particle were required for ignition of the target with the possible exception of aluminum targets. Further work is planned to expand the range of operating conditions with the ultimate aim of developing a model of the ignition process. Completion of this program will throw light on the mechanisms of ignition by particle impact and help to define measures to protect against this hazard.

Benz and Stoltzfus reported on the ignition of metals by frictional heating. In this test, the end of a hollow cylinder is rotated against the end of a coaxial, stationary, hollow cylinder of the same dimensions. The contact pressure P and the relative linear velocity v between the cylinders are measured. The product Pv (a measure of power applied per unit area) to reach ignition is used to rank the materials tested, with higher values indicating greater resistance to ignition. Of the materials tested, alloys high in nickel and copper required higher Pv for ignition than alloys high in iron. Aluminum and titanium required the lowest Pv for ignition. Additional work is planned to establish the effects of system variables on the Pv values for ignition, with the aim of developing an ignition model. Completion of this program should provide valuable information for use in the selection of materials for rotating machinery and other devices where significant rubbing can occur between components.

Yuen presents a theoretical model of the metal ignition process, which includes the effects of those parameters known to be important in metal ignition. Based on numerical solutions of the model, three cases of metal ignition control are identified. The concept of minimum ignition heat flux is introduced and is suggested as a more appropriate parameter for ranking resistance to ignition than the traditional ignition temperature. Qualitative comparisons are made between predicted ignition conditions and those measured by Benz and Stoltzfus for ignition by frictional heating. Comparisons with other measured ignition conditions and further refinement of the model should eventually provide valuable insights into the ignition process. This should, in the future, improve the ability of designers to minimize the potential for ignition in oxygen handling systems.

Bransford reports on his development of a laser heating technique to study the ignition of metals in high-pressure oxygen. For aluminum, which burns in the vapor phase, he concludes that ignition requires the exposure of fresh metal to either reactive, reducible oxides of alloying elements or directly to oxygen by rupture of the protective aluminum oxide layer. For surface burning metals, he concludes that ignition is initiated by sudden exposure of reactive material to the hot oxygen environment, decomposition of certain oxides, or phase changes in the oxide layer. Continued development of this technique should help improve our understanding of the ignition process and, hence, our capability to control it. Comparisons of the experimental data with the predictions of Yuen's ignition model would seem to be appropriate.

Ignition of Nonmetals

McIlroy and Zawierucha introduce the use of a new instrument, the accelerating rate calorimeter (ARC), as a potentially useful device to evaluate nonmetals for use in oxygen service. This device can detect self heating of a test material at very low rates and track the rising temperature to a point of runaway. For six materials, the authors used an ARC to measure the temperature at which the maximum heating rate occurred in oxygen at pressures to 37 MPa (5400 psi). These temperatures are compared with the standard autogenous ignition temperature (AIT) measured according to ASTM Test Method for Autogenous Ignition Temperature of Liquids and Solids in a High-Pressure Oxygen-Enriched Environment (G 72), and were found to be within about 30 K. With the exception of a hydrocarbon oil at low oxygen pressure, self heating was not detected at a temperature more than 100 K less than the AIT. Testing of a wide variety of additional materials will be required to establish the benefits of ignition data as measured by the ARC in comparison to the simpler AIT measurement.

Another instrument which seems to have potential value for testing of materials for oxygen service is the pressurized differential scanning calorimeter (PDSC). Bryan and Lowrie report comparative data for four materials whose ignition temperatures were measured by ASTM G 72 and by PDSC in oxygen at pressures to 6.8 MPa (1000 psi). These data show that the ignition temperature measured by PDSC is significantly higher than the standard AIT. The difference between the two values does not seem to be consistent. It must be concluded, therefore, that considerable additional work needs to be done to understand the differences and to establish the value of testing by PDSC. As with the ARC, the PDSC can identify incipient reactions below the AIT, which may be of significant interest in some applications.

Three papers report data related to the propagation of combustion of metals. Two of these concern propagation rates of several preheated metals in high-pressure oxygen. The third reports the pressurized oxygen index for carbon steel. The study of propagation is important to the understanding of potential damage that can occur if a fire is started. It can also help to define conditions that will not allow combustion, in which case damage can be essentially negligible even if an ignition occurs.

Sato and Hirano report the fire spread rates for aluminum and steel rods (3 mm diameter) immersed in oxygen at pressures to 10 MPa (1450 psi) after preheating in argon to temperatures as high as 873 K for aluminum and 1573 K for steel. The effect of surface oxide coatings on each metal and of gold plating on steel was investigated. The results are discussed in terms of a model developed from previous work. Some of the complex effects that are seen are attributed to changes in the contour of the solid surface at the interface with the molten, burning droplet. While the data presented add greatly to our understanding of the combustion process, and extend previous referenced data published by the authors, many additional factors must be investigated before it can be applied to routine selection of materials for oxygen service. For example, data for other metals is needed, and the effects of sample geometry should be investigated.

Benz, Shaw, and Homa report data from experiments to measure burn propagation rates of 3.2-mm-diameter rods of four metals and alloys in oxygen at pressures to 68.9 MPa (10 000 psi). With the specimens initially at room temperature, nickel and copper could not be ignited and Monel 400 was only partially burned. Rates for 304 and 316 stainless steels and Inconel 718 were similar and increased with oxygen pressure. Rates for aluminum were an order of magnitude higher. With an initial specimen temperature of about 900 K, nickel was not ignited, Monel again burned only partially and the burn rate of 316 stainless steel was increased. As with Sato and Hirano's work, it will be useful to extend these data to cover other metals and to test other sample sizes and geometries such as tubes and strips.

The standard oxygen index test (ASTM Method for Measuring the Minimum Oxygen Concentration to Support Candle-like Combustion of Plastics (Oxygen Index) [D 2863]) for measurement of the flammability of nonmetals has been extended by Benning and Werley for use with carbon steel at pressures to 20 MPa (3000 psi). Tubular specimens, 25 mm (1 in.) in diameter by

4.8 mm (0.19 in.) wall were ignited by a thermite pill in flowing oxygen diluted by nitrogen. The reported data show that, for the geometry used, steel was just flammable in 81% oxygen at 1 MPa, 53% at 7 MPa and 51% at 20 MPa. It was also shown that Monel was a suitable material to use for a fire-break in a steel pipe. The use of this test, perhaps in a simpler form, to obtain data for other metals could provide an additional ranking method for selection of metals for oxygen service. Of even greater significance, it could lend valuable insight into the combustion processes for metals and expand our capability for controlling them.

System Cleanliness and Contamination

Five of the papers presented covered the related topics of cleanliness and contamination. The first of these reports the dramatic and costly results of contamination in combination with component deterioration, choice of materials, and improper operation. Another paper reviews an investigation of cleaning methods and procedures used by the military. The next two papers present a comparison of four commonly used methods for the detection of oil film contamination, and an instrumental method for continuous monitoring of oil leakage in reciprocating oxygen compressors. The final paper proposes a new method for simple determination of the quantity of oil residue in solvents that may have been used for component cleaning or extraction.

Grubb reports on the destruction of a $6 000 000 aircraft in a ground fire incident following the removal of an on board oxygen cylinder. The fire is thought to have started in a corroded, aluminum poppet valve with a silicone seat that allowed leakage of oxygen that had not been bled from the system as procedures required. In addition, poor filtration in the oxygen filling line had allowed widespread contamination of the whole piping system with iron oxide from the ground filling cylinders. This paper provides a valuable case study, which illustrates the importance of attention to all aspects of the design and use of oxygen systems. It will bring home, to both the novice and the experienced, the essential need for careful selection of materials, the need for cleanliness throughout the life of a system, and the need for strict adherence to proper procedures.

Schroll presents a detailed report of requirements by the military for cleaning of equipment for use in oxygen service and for cleaning aircraft oxygen systems. The report is based on responses by industry and the military to a survey made in 1983. Included are clean room specifications, cleaning procedures, and inspection and packaging requirements for equipment. Cleaning procedures for specific equipment and military and industrial cleaning specifications are tabulated. Detailed procedures for flushing and purging aircraft oxygen systems are included. The standards and procedures discussed are referenced. This paper will provide an excellent reference and source document for those concerned with many aspects of cleaning for oxygen service. It also serves to highlight some of the gaps in our knowledge and capabilities in this critical area, particularly in defining and measuring levels of cleanliness, a topic which is also addressed in the next paper.

Gilbertson and Lowrie present the results of measurements of the sensitivities of four tests commonly used to detect oil contamination on surfaces. Of the four tests, only the water break test was able to reliably detect all six hydrocarbon oils used at levels below 200 mg/m² (19 mg/ft²), about the highest concentration accepted in industry as an adequate cleanliness level. Unfortunately, the water break test is the most difficult of the four to apply since it requires smooth horizontal surfaces to work. Also, it should be noted that none of the oils tested was water soluble, as many cutting oils are. The results of these tests call into serious question the adequacy of these techniques widely used to qualify the cleanliness of parts and systems used in oxygen service. This is an area where further, extensive study is greatly needed.

Ernst describes a new detection system for the continuous monitoring of reciprocating oxygen compressors for oil leakage. The oil detector can be attached to the distance pieces that separate the lubricated compressor drive from the pistons. The detector uses the dielectric effect of oil on a capacitor to detect any leakage. Extensive field tests are cited regarding the reliability of the system which will be a valuable addition to the available techniques for the safe operation of oxygen compressors.

In the final paper, Werley proposes a simple and inexpensive technique for the determination of the concentration of nonvolatile residue in solvents. To make the measurement, a sample of the solution is carefully introduced into the capillary portion of a sealed, disposable, Pasteur pipet. Proper evaporation of the solvent allows the residue to form a column in the capillary whose length is proportional to the volume of residue in the sample. If fully developed, the technique should have application in several areas relating to cleaning for oxygen service. For example, it could be used to measure the quantity of oil in solvent used to wash cleaned equipment to determine the level of residual contamination.

At this point it is appropriate to gratefully acknowledge the contributions of John O. Cronk for acting as co-chairman for the symposium and Robert Waller and Drew Azzara for liaison with ASTM and their help in planning and organizing the symposium.

Michael A. Benning

Corporate Research Services Department, Air Products and Chemicals, Inc., Allentown, PA, 18105; symposium co-chairman and editor.

John A. Gilbertson[1]

The Application of Compatibility Knowledge

REFERENCE: Gilbertson, J. A. **"The Application of Compatibility Knowledge,"** *Flammability and Sensitivity of Materials in Oxygen-Enriched Atmospheres: Second Volume, ASTM STP 910*, M. A. Benning, Ed., American Society for Testing and Materials, Philadelphia, 1986, pp. 7–15.

ABSTRACT: The author explains how experience and empirical rules are used in the design of equipment for use in oxygen-enriched atmospheres. Standard methods of testing materials are making the availability of consistent data more widespread. The lack of an agreed method of applying those data founded on theory and science impedes progress and is a barrier to educating newcomers.

KEY WORDS: oxygen, standards, education, liquid tests, gaseous tests, codes of practice

I felt greatly honored to have been asked by the symposium committee to give the keynote address to the second symposium on Flammability and Sensitivity of Materials in Oxygen Enriched Atmospheres. ASTM Committee G-4 on Compatibility and Sensitivity of Materials in Oxygen Enriched Atmospheres must be congratulated for the work it has done during the last decade in identifying the components of this problem area and then systematically developing test methods, experience data, standards, and guidelines for use by qualified personnel in the selection of material for oxygen service. Note the use of the term "qualified personnel" and reflect that "flammability and sensitivity of materials" is not currently in general capable of being taught from basics as no unchallengeable theories exist for the universal education of the tyro. I will return to this matter later.

You will forgive me, I am sure, if I take this opportunity of looking backwards to a point before Committee G-4 had been formed. In 1973 Dr. A. Lapin of Air Products and Chemicals Inc. read a paper [1] at the Interna-

[1]Group technical manager, The BOC Group, Hammersmith House, London, England W6 9DX.

tional Institute of Refrigeration held in Brighton, England, and introduced to a wide audience the concept of an acceptability index based on seven properties of a material. Because of the complexity of that index, a simpler arbitrary relationship was used in practice

$$\text{acceptability index} = (OI)^2(T/C)$$

where

OI = oxygen index expressed as the minimum percentage of oxygen in a mixture of oxygen and nitrogen required to sustain candle-like burning,

T = auto-ignition temperature, K, in 100% oxygen at 1 atm, and

C = heat of combustion, MJ/kg.

Furthermore, Lapin proposed that two materials with equal acceptability indexes would have equivalent oxygen compatibility qualities. Whether or not one agreed with the proposition was secondary at that time to the fact that a debate was catalyzed with far reaching effects. Shortly after, Lapin and I found ourselves working together in Paris with time to discuss ways and means of developing an industry wide exchange of experience and views that would, we hoped, lead to standardized methods of testing and selection. For a while we were strongly of the opinion that the work could best be done in Europe where the writing of industrial codes was then well developed and relatively quick, but finally we decided that the resources and wider interests of the United States made it the more appropriate country to commence the meetings.

In 1974, ASTM Committee G-4 was inaugurated, and membership has included national and international oxygen producers, armed services, industry, National Aeronautics and Space Administration (NASA), and insurance companies. Its relevance and competence is illustrated by the fact that no competing or parallel committee exists in Europe. Our original decision was right.

Why Are We Concerned?

The use of oxygen incurs risk, and our role is to minimize that risk. During this century, judgment has been applied and experience gained by oxygen users on what materials can be used. Normal engineering materials were shown not to be satisfactory unless precautions were taken, and it is not difficult to find examples of mild steel pipelines burning through because of high gas velocity and low levels of cleanliness. Similarly aluminum alloy pumps, once commonplace, were restricted in use for high-pressure high-speed duties following a series of spectacular fires in the late 1960s.

The growing use of oxygen since the second world war was exposing the unsatisfactory position of the industry in solving compatibility problems, and considerable efforts were made in many countries to correct the situation.

The usual fire prevention technique of removing one leg of the fire triangle is made difficult because, by the nature of the problem, oxygen is present and many materials readily burn in it. Slusser and Miller [2] in their paper, "Selection of Metals for Gaseous Oxygen Service," produced an extended fire triangle that is excellent in clarifying the boundaries of the problem. They point out that

• all materials are potential fuels, and the system design must ensure that ignition of a lower compatibility material does not kindle reactions in higher compatibility material;
• the system must accommodate a specific oxygen environment, and temperature, pressure, velocity and quantity can affect compatibility;
• ignition is the least understood aspect of the combustion process, and all possible causes should be considered.

Our concern, therefore, derives from the fact that most materials are combustible in oxygen under some conditions. As it is not possible to eliminate all sources of ignition, it is important to be able to select the most appropriate materials and design them into the system so that the probability of ignition is reduced and damage minimized. At the same time we live in a practical world and solutions that are not economic may well be unacceptable.

What Is Done?

Unlike electrical resistance or ultimate tensile strength in metals or compressive strength of structural concrete, which we can use in design, there is no obvious and incontrovertible property that we can measure and use with confidence to ensure oxygen compatibility in equipment. That is not to say that measurements are not made; quite the contrary, as laboratories around the world measure ignition temperature, heat of combustion, promoted ignition, liquid oxygen impact, and so on to produce their own lists, produce their own lists of materials ranked in order of risk. We know only too well that these orders of ranking of risks are dependent not only on the material tested, its size, and its geometry but on the test equipment used.

Until the recent advances in providing data, we lived with the paradox of different organizations producing and using their own data in a manner different from other organizations without noticeable differences in their design failures. Engineers have, without the support of an underlying science, continued to produce improved versions of machines that work satisfactorily in oxygen service. The latest model of a machine has sometimes also been the test-bed model disappearing in a flash as a new failure mechanism is met for the first time.

The application of quasi-scientific knowledge is not unknown in engineering, but the result can be uneconomic overdesign as the factor of safety (or in that case, coefficient of ignorance) cannot be determined.

The dilemma that we face was brought home to me last year when I was giving a lecture on design in oxygen systems to undergraduate engineers. The lecture concluded with the description of a disciplined approach to problem solving. Attendees to this symposium would be familiar with the method outlined. The method is as follows:

(a) Define duty.
(b) Choose materials with suitable mechanical properties.
(c) Consider ignition mechanisms and probabilities.
(d) Design out/minimize risk.
(e) Study consequences of failure.
(f) Decide what factor of safety is appropriate.
(g) Make choice.

What at first sight is plausible turns out not to be so, as it is not possible to take Step g without having experience and knowledge and being able to apply judgment. In fact, I was describing a method for the initiated to use and was not educating the uninitiated.

With some notable exceptions much material choice is based on successful experience in similar circumstances. This method allows industry to make cautious advances without fully understanding its successes or failures. The method also leads to unsolvable arguments about the suitability of materials when the experience of one protagonist runs counter to the experience of another. I remember the suitability of stainless steel in oxygen systems being argued with the fervour of religious bigots by opposing sections of an international committee.

The Search

This symposium is an example of the cooperation that exists among organizations in the search for better understanding of ignition mechanisms and means to avoid them. The work that is done falls roughly into three overlapping categories of study.

1. Codes of Practice and Guidelines of Behavior. A most fruitful area of cooperation is the sharing of experiences, good or bad, to seek agreement on recurring reasons for incidents and minimum standards of behavior for avoiding a repetition. In Europe we have been very successful in coming to reasonable speedy agreements on a variety of subjects concerning oxygen usage. The subjects chosen reflect past problems in industry. These documents are good reference documents for purchasing, layout, operation, and maintenance. They cannot do the work of the designer although they often state preferred design.

As examples, I quote documents written by the Industrial Gases Committee (IGC) [3], which are on their current publication list:

- 8/76/E—Prevention of accidents arising from enrichment or deficiency of the oxygen in the atmosphere.
- 12/80/E—Pipelines distributing gases and vacuum services to medical laboratories.
- 10/81/E—Reciprocating compressors for oxygen service.
- 11/82/E—Code of Practice for design and operation of centrifugal liquid oxygen pumps.
- 13/82/E—The transportation and distribution of oxygen by pipeline.
- 27/82/E—Turbo Compressors for oxygen service (evolved from the work of the European Oxygen Turbo Compressor Working Panel, formed May 1970).
- 20/83/E—Distribution of oxygen, acetylene and methylacetylene mixtures at users works.

2. The pooling of experience is not always the most appropriate way of finding an answer to an existing problem, and it may be irrelevant if a significant departure from existing practice is envisaged. In these circumstances, highly specific tests are used in which test conditions attempt to mimic the expected configuration of the part and its failure condition. Examples of specific testing are as follows:

Wegener W., "Investigations on the Safe Velocity of Flow to be Admitted for Oxygen in Steel Pipe Lines," *Stahl Eisen*, Vol. 84, No. 8, 9 April 1964, pp. 469–475.

This work was done because the existing limiting allowable velocity in steel pipes had been 8 m/s since 1942, and by the 1960s it was thought to be too low. A series of tests were made in which solids like rust, dust, sand, mill scale, welding slag, coke, coal, and iron powder were introduced into the oxygen at various speeds in an endeavor to determine the conditions required to start fires in steel pipelines. The workers concluded that there were no objections to using velocities higher than 8 m/s in pipelines as in practice uneconomic pressure drops would set the limit; however, the hazard to valves and fittings would be more severe.

Bauer, H., Klein, G., Wegener, W., and Windgassen, K., "Fire Tests on Centrifugal Pumps Handling Liquid Oxygen," *Cryogenics*, Dec. 1971, pp. 469–476.

A series of tests were run aimed at clarifying the conditions under which fires or explosions may be induced in centrifugal pumps. Heavy rubbing was arranged, and separately, foreign bodies were introduced into the specimen pumps. The results were useful subsequently in proposing suitable measures for improving safety.

Naegeli, J. P., "What Triggers Oxygen Turbo Compressor Fires?," Oxygen Compressors and Pumps Symposium CGA, Atlanta, GA, 1971, pp. 35–38.

The tests, made in oxygen at 50 bar and 170°C, showed the effect of rubbing labyrinth tips at 210 m/s against stationary bushing materials. One objective was to find material combinations that would not ignite under those severe conditions.

Jenny, R. and Wyssmann, H., Friction-Induced Ignition in Oxygen, *Flammability & Sensitivity of Materials in Oxygen-Enriched Atmospheres, STP 812*, American Society for Testing and Materials, Philadelphia, 1983, pp. 150–166.

Oxygen pressures as high as 120 bar call for a quantitative approach to the safety problem for high pressure compressors. A proposed theoretical model of ignition induced by rubbing was confirmed by the experiments made.

Stradling J., Pippen D., and Frye G., "Techniques Employed by NASA WSTF to Ensure Oxygen System Component Safety," *Flammability & Sensitivity of Materials in Oxygen-Enriched Atmospheres, STP 812*, American Society for Testing and Materials, Philadelphia, 1983, pp. 97–107.

At high pressures and flow rates, components may fail if designed to conventional rules. Components are subjected to a configuration test providing an environment at least as severe as the end use. Upgrading of materials and cleaning procedures may be required if ignition is to be avoided.

The results of these tests are very useful although they are qualified by the understanding that while an ignition in test is a strong warning, a nonignition in tests does not guarantee a nonignition in practice. It is also difficult to transfer the results of such a specific test to other situations.

3. The third category of study is the collection of data through tests of specific properties, like auto-ignition temperatures and the use of them to test ignition hypotheses. Much of that work is done in the United States. ASTM Committee G-4 is fortunate to have many of the experts who are collecting and interpreting that information on its committees.

From a mainly empirical position, the search for better understanding of sensitivity and flammability continues, and as time passes, predictions of safe operation of materials in oxygen enriched atmospheres can be made with greater confidence.

The Selection

The objective of workers in this field is to select the most practical and acceptable material for duty in the given oxygen-enriched environment and to continue to use it successfully through a profitable life. Not surprisingly, the choices available reflect the means available for the study of the ignition mechanisms.

1. When no change of duty is expected, the decision to make no alteration in materials used is a conscious option based on previous successful experience in the same duty. Checks that the material supplied is as previously used may be necessary.

2. Small change from previous experience. When existing materials are to be substituted by different materials with no change of duty or where existing materials are to be used in a modified duty, a paper study by the discipline described in ASTM Guide for Evaluating Nonmetallic Materials for Oxygen Service (G 63) is recommended to ensure that safety margins are not eroded.

3. Entry into a known field. Reference to Codes of Practice and similar documents would be a first step. For example, CGA pamphlet G-4.4 summarizes industrial practices used in gaseous oxygen transmission and distribution piping systems such as encountered in chemical works, refineries and so forth. The pamphlet clearly states that it is not a design handbook, but it is a useful source of materials that have a wide acceptance based on successful use.

4. Substantial step from previous experience. Testing at some level will almost certainly be called for. Where the deviation from experience concerns material used, tests will be made to compare its properties with known materials. Where the duty or design configuration is unusual, tests similar to those described earlier (Item 2, p. 11) will be needed.

Assistance is also available from the "Design Guide for High Pressure Oxygen Systems," NASA Reference Publication 1113, Aug. 1983, written to provide information on "the design subtleties, techniques and related knowledge, presently in use in aerospace applications." The authors compiled the document specifically to provide a ready and easy reference for designers of any high-pressure system. Another relevant publication is ASTM Guide for Designing Systems for Oxygen Service (G 88).

As the designer moves further away from past experience the relationship between the test data and what is acceptable for the proposed duty becomes more important. Care must be exercised in evaluating test data, as the end usage may inflict on the materials quite different conditions from those of the test method.

At the present time there is little alternative but to accept arbitrary relationships in making the transition from candidate material test data to the acceptability of the candidate for the proposed duty. Neary [4] showed that the most commonly used test methods were ignition temperature, heat of combustion, and mechanical impact (in liquid and in gaseous oxygen). Data obtained with precision from these tests are used qualitatively in framing rules to control material selection.

• Ignition temperatures. ASTM G 63 prefers ignition temperatures of above 440°C, warns against temperatures below 160°C, and suggests a temperature 100°C higher than the maximum use temperatures. In the past when the BOC Group measured ignition temperatures at 20 bar compared to ASTM 100 bar it required the candidate material ignition temperature to be 150°C in excess of the maximum use temperature. The BOC Group now favors the ASTM standard.

• Heat of combustion. Lowrie [5] reports that materials with heat of combustion over 41.9 MJ/kg are potentially dangerous for most applications in oxygen enriched atmospheres and that a small amount of nonmetallic oxidizable material with a heat of combustion below 6.3 MJ/kg, if ignited, is unlikely to ignite adjacent metal parts.

• Mechanical impact. This test is a simple method of transmitting a measurable amount of energy to the sample. NASA Reference Publication 1113 reports that no component ignitions have been attributable to materials that have passed the appropriate test.

The Future

The individual papers of this symposium illustrate quite clearly that much is being done to improve our knowledge of flammability and sensitivity of materials in enriched atmospheres. Three quarters of the papers are concerned with collecting data on various methods of ignition. The total amount of data now available is very large, although it is not universally available and it is not consistent.

We must encourage ASTM Committee G-4 to continue with their good work, to produce test methods and to press for their adoption internationally. They should also become the repository of all consistent test data so that parallel testing will become unnecessary. Material specification checks will still be necessary at location.

The question of how best to apply the material data into new design configurations still has to be solved. More attention must be given to the application of the knowledge we now have, if you agree with me that arbitrary rules are not the way forward. We cannot expect innovation, invention, and improvement to flow freely in an area lacking an agreed theoretical and scientific foundation. These deficiencies cause problems in education too, and I ask you to consider how young engineers or scientists become skilled in these matters. Where are the books to study or the courses to attend? The acquisition of knowledge is a critical part of being effective in developing reliable, safe equipment, and procedures, and it is being handled in a piecemeal and ineffective way.

Training quite clearly is done well by some organizations, but I criticize our failure to educate. I recommend that ASTM make this an issue for the third symposium on flammability and sensitivity of materials in oxygen-enriched atmospheres. The debate, that such action will encourage, could clarify the definition of "qualified personnel" and the means of achieving the qualification.

In concluding this address I remind you that cooperation of the last decade has not only shown us how much we know but also exposed what we still have to learn. Let us not fail to fill the gaps in our knowledge.

References

[1] Lapin, A., "Oxygen Compatibility of Materials," International Institute of Refrigeration Commission Meeting, Brighton, England, 1973.
[2] Slusser, J. W. and Miller, K. A., "Selection of Metals for Gaseous Oxygen Service," *Flammability and Sensitivity of Materials in Oxygen-Enriched Atmospheres, STP 812*, American Society for Testing and Materials, Philadelphia, 1983, pp. 167-191.
[3] Industrial Gases Committee, Publications de la Soudure Autogene, 32 Boulevarde de la Chapelle, Paris, France.
[4] Neary, R. M., "ASTM G 63: A Milestone in a 60-year Safety Effort," *Flammability and Sensitivity of Materials in Oxygen-Enriched Atmospheres, STP 812*, American Society for Testing and Materials, Philadelphia, 1983, pp. 3-8.
[5] Lowrie, R., "Heat of Combustion and Oxygen Compatibility," *Flammability and Sensitivity of Materials in Oxygen-Enriched Atmospheres, STP 812*, American Society for Testing and Materials, Philadelphia, 1983, pp. 84-96.

Frank J. Benz,[1] Ralph E. Williams,[2] and Dan Armstrong[2]

Ignition of Metals and Alloys by High-Velocity Particles

REFERENCE: Benz, F. J., Williams, R. E., and Armstrong, D., **"Ignition of Metals and Alloys By High-Velocity Particles,"** *Flammability and Sensitivity of Materials in Oxygen-Enriched Atmospheres: Second Volume, ASTM STP 910*, M. A. Benning, Ed., American Society for Testing and Materials, Philadelphia, 1986, pp. 16–37.

ABSTRACT: The ignition of metals and alloys by impacting high-velocity particles in gaseous oxygen was investigated. A convergent/divergent nozzle was used to accelerate the flowing oxygen, which in-turn accelerated the particles to velocities greater than 305 m/s (1000 ft/s). The test sample (target) was placed at the end of the chamber in the flow path.

Aluminum 6061, Type 316 stainless steel, Type 304 stainless steel, and Inconel 718 were ignited with 1600-μm aluminum 2017 particles at elevated temperatures and pressures whereas Monel 400 could not be ignited. The ignition susceptibility of metals and alloys appeared to increase as the inlet pressure, sample temperature, and particle size were increased. Type 304 stainless steel particles required more extreme conditions for ignition of materials than similar sized aluminum particles. The results indicated that ignition and subsequent burning of the particles were required for ignition of the target materials with the possible exception of Aluminum 6061.

KEY WORDS: ignition, ignition criteria, ranking, aluminum, stainless steels, particle impact, oxygen compatibility, particle impact, Inconel, Monel

The impact of high-velocity metallic particles on metal surfaces has been suspected for many years to be the cause of oxygen fires in gaseous oxygen systems [Refs *1–7* and ASTM Guide for Evaluating Materials for Oxygen Service (G 63)]. However, testing to characterize the major parameters involved in this ignition process has been limited. The National Aeronautics and Space Administration (NASA) White Sands Test Facility (WSTF) is presently involved in developing several test methods for evaluating metals and alloys suitable for oxygen service. One of the methods consists of an ap-

---

[1]Materials engineer, NASA, Johnson Space Center, White Sands Test Facility, Laboratories Test Office, Las Cruces, NM 88004.

[2]Senior engineers, Lockheed/EMSCO, Johnson Space Center, White Sands Test Facility, Las Cruces, NM 88004.

paratus for impacting high-velocity particles on target materials. This paper will describe the apparatus and present data that have been obtained to date.

Background

Review of Previous Testing

The ignition of metal particles in oxygen type atmospheres has been studied by many investigators [8-15]. However, very little work has been performed on the ignition characteristic of impacting metal particles on other metal or alloy surfaces. Two of the earliest studies on this subject were reported in the early 1960s by Nihart and Smith [16] and Wegener [4].

Nihart and Smith [16] reviewed testing conducted by the Union Carbide, Linde Division, where known quantities of powder materials were accelerated in high-velocity oxygen streams and impacted against metal specimens. The size and type of particles and the velocity of the oxygen were not specified. However, it is believed that the size of the particles was less than 75 μm, and the gas velocity was subsonic. The following results were reported:

1. Carbon steel, cast iron, stainless steel, and aluminum could be ignited and completely consumed with the evolution of large quantities of energy.

2. Stainless steel and aluminum burned with explosive violence.

3. Copper, copper alloys, and Monels had a decided quenching effect on combustion.

Wegener [4] conducted tests in which various types of contaminant (less than 5000 μm) were entrained into flowing oxygen in straight and curved steel lines at 2.8 MPa (400 psig). Approximately 1 to 2 kg (2.2 to 4.4 lb) of contaminant was injected over a time period of 30 to 60 s. The velocity limits for ignition of the steel lines were determined (Table 1).

More recently, Phillips [17] reported that ignition of Type 304 stainless steel tubes occurs when aluminum fibers were ignited in oxygen by dynamic oscillation in a resonance tube.

Monroe et al [18] conducted tests in which single 1000-μm (0.039-in.) silica particle was accelerated by a particle gun to velocities of 427 m/s (1400 ft/s) and impacted on various materials. The oxygen pressure was nominally 6.9 MPa (1000 psig), and the sample materials were preheated up to a maximum temperature of 1120 K (1550°F). Typical materials tested by these investigators are listed in Table 2. One ignition was observed with American Iron and Steel Institute (AISI) 4140 steel, which had been preheated to 911 K (1180°F). However, this ignition could not be repeated even at sample temperatures up to 1120 K (1550°F). No other ignitions were observed with any of the other test materials listed in Table 2.

TABLE 1—*Ignition velocities of entrained particles in oxygen (steel pipe)* [4].

Pipe Line Configuration	Entrained Material	Nonignition Velocity, m/s	Ignition Velocity, m/s
Straight	rust	84[a]	...
Straight	sand	84[a]	...
Straight	steel scale	28	52
Straight	weld slag	44[b]	...
Curved	weld slag	53[c]	...
Straight	coke	30	...
Curved	coke	17	53
Curved	carbon	...	34
Straight	carbon	...	13
Straight	80% sand, 20% iron powder	13	31

[a]Maximum test velocity attainable.
[b]Particles began to glow at this velocity.
[c]Some sparks at this velocity.

TABLE 2—*Materials tested by Monroe et al* [18].

1. Carbon steel, AISI 1025
2. AISI 4140 steel
3. Ductile iron
4. 304 stainless steel
5. 17-4 pH stainless steel
6. 410 stainless steel
7. Lead babbitt
8. Tin babbitt
9. Inconel 718
10. Aluminum 1100

Previous Testing at WSTF

Beginning in the late 1970s and early 1980s, a concentrated effort at the NASA WSTF was initiated to investigate the conditions that would produce ignition of metals and alloys caused by high-velocity particles. The emphasis for this study was based on testing conducted by Porter [19] in which a single 800-m (0.032-in.) aluminum 2219 particle injected into a flowing stream of oxygen at 31 MPa (4500 psig), 277 K (530°F), and 1.14 kg/s (2.5 lb/s) ignited and totally consumed an oxygen flow control valve. Repeating tests on a similar valve at the same conditions but injecting 0.15 g (3.3 × 10^{-4} lb) of 150-m (0.0059-in.) aluminum 2024 or Inconel 718 particles did not ignite the valve.

Based on the above results, a particle impact chamber was developed (Fig. 1) for the purpose of evaluating the ignition susceptibility of materials to particle impact [19]. The tests were conducted by accelerating particles through a convergent/divergent orifice (velocities greater than Mach 1) and allowing the particles to impact on various target materials (1.9 cm diameter, 0.25 cm

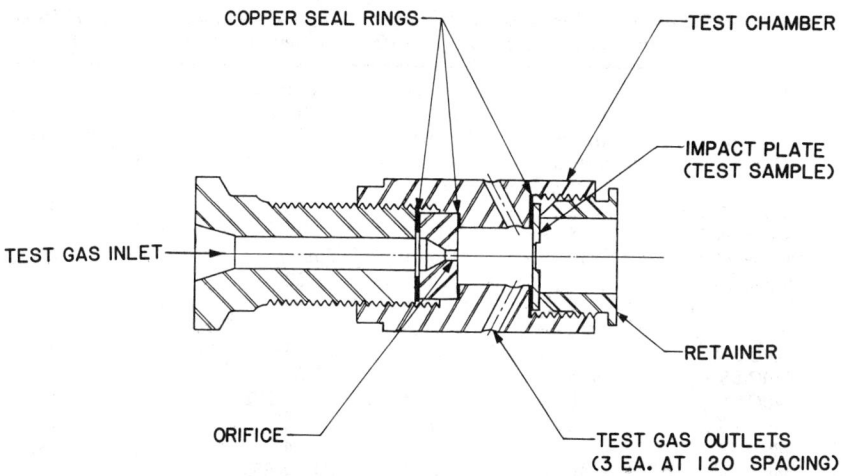

FIG. 1—*First generation sonic velocity impact chamber.*

thick) at elevated temperature and pressures. Tables 3 and 4 give the results for injection of 800- and 150-μm (0.031- and 0.0059 in.) aluminum particles at supersonic gas velocities. In the case of injecting 800-μm aluminum particles, ignitions were observed with Inconel 718, 440A stainless steel, A286 stainless steel, 21-6-9 stainless steel, and 304 stainless steel, whereas no ignitions were observed with Monel 400 and 440C stainless steel. When 0.15 g (3.3 × 10^{-4} lb) of 150-μm aluminum particles were injected, 21-6-9 stainless steel was the only material that produced ignitions. Tests were also performed injecting 150- and 800-μm Inconel 718 particles at conditions similar to those conducted using aluminum particles. Materials that produced ignitions with the aluminum particles did not ignite when subjected impact of Inconel 718 particles. In several tests, the orifice was removed from the chamber and orifices were placed in the gas exit lines. These modifications produced subsonic gas flow, and tests were repeated at similar initial pressures and temperatures as used for the supersonic tests. The alloys that ignited at the supersonic gas velocities did not ignite at the subsonic velocities.

More recently, testing was conducted on several metals and alloys using the same supersonic particle impact chamber (Fig. 1) to determine the inlet oxygen temperature required for ignition; the results were reported by Schoenman [20]. In these tests, 1600-μm (0.063-in.) aluminum 2017 particles were injected into a flowing oxygen stream at an initial pressure of 31 MPa (4500 psig) and an oxygen flow rate of approximately 0.45 kg/s (1.0 lb/s). The results are given in Table 5. As observed in previous tests, materials high in nickel and copper did not ignite whereas materials high in iron did ignite. In the case of the alloys that did ignite, Hastelloy X required the highest temper-

TABLE 3—*Results of particle impact tests using 800-µm (0.032-in.) aluminum particles at flow rates of 0.45 kg/s (1.0 lb/s); sonic velocities* [19].

Test Sample Material	Number of Particles	Average Inlet Pressure, MPa	Average Inlet Temperature, K	Results	
				Number of Ignition	Number of Tests
Monel 400	2	32	550	0	1
Monel 400	5	32	551	0	3
Monel 400	10	30 to 31	561 to 578	0	5
Inconel 718	2	28	529	0	1
Inconel 718	5	27 to 28	583	0	2
Inconel 718	10	27 to 28	533 to 569	1	3
440C SS[a]	2	31	555	0	1
440C SS	5	31	561	0	1
440C SS	10	29 to 31	561 to 572	0	5
440A SS	2	32	555	0	2
440A SS	5	32	544	0	1
440A SS	10	31 to 32	544 to 555	1	2
A286 SS	2	31 to 32	533 to 566	0	2
A286 SS	5	31	550 to 569	2	3
A286 SS	10	31	572	1	1
21-6-9 SS	2	32	538 to 555	0	2
21-6-9 SS	5	31	566	0	2
21-6-9 SS	10	31 to 37	539 to 566	1	2
304L SS	1	22 to 24	544	2	2

[a]SS is stainless steel.

TABLE 4—*Results of particle impact testing using 0.15 g (3.3 × 10⁻⁴lb) of 150-µm (0.0059-in.) aluminum particles at flow rates of 0.45 kg/s (1.0 lb/s); sonic velocities* [19].

Test Sample Material	Average Inlet Pressure, MPa	Average Inlet Temperature, K	Results	
			Number of Ignition	Number of Tests
Inconel 718	27 to 29	522 to 553	0	20
440A stainless steel	29 to 33	539 to 566	0	20
A286 stainless steel	29 to 35	511 to 594	0	20
Stellite	28 to 32	541 to 593	0	20
21-6-9 stainless steel	24 to 33	539 to 595	2	22
Copper 102	28 to 35	542 to 660	0	40
Copper beryllium	28 to 34	505 to 553	0	20

ature for ignition whereas Type 316 stainless steel required the lowest temperature for ignition.

Development of Standard Test Method

In 1982, the Office of Chief Engineer, NASA Headquarters, Washington, DC, tasked NASA WSTF to develop a method capable of evaluating materials for oxygen service. Since it was demonstrated that the impact of high-

TABLE 5—*Results of particles impact tests using ten 1600-μm aluminum particles at initial oxygen pressure of 31 MPa and flow rate of 0.45 kg/s* [20].

Test Sample Material	Lowest Ignition Temperature, K	Results Maximum Temperature Nonignitions, K
Zirconium copper alloy	. . .	694
Nickel 200	. . .	713
K Monel 500	. . .	713
Monel 400	. . .	700
Silicon carbide	. . .	711[a]
Hastelloy X	658	. . .
Invar 36	644	. . .
316 stainless steel	505	. . .

[a]Material shattered on impact.

velocity particles could ignite many materials, a new improved chamber was developed based on the experience gained from the previous testing. The general concept of a supersonic velocity impact tester was retained but redesigned so that the conditions of the flowing oxygen, the particles, and the target material before impact could be determined.

The new impact chamber has been built, and testing to evaluate its capabilities to determine the ignition characteristics of metals by impacting high-velocity particles has been initiated. The following sections in this paper will review the testing to date using this new impact chamber. Test data will be presented to describe the effects of varying initial pressure and temperature, and size and type of particles. Conditions required for ignition of several different materials will be compared.

Test System Description

The modified supersonic velocity particle impact chamber (Fig. 2) consisted of three major sections: (1) gas inlet and flow straightener, (2) particle injector and converging nozzle, and (3) diverging nozzle and test sample holder. Gaseous oxygen (GOX) entered the chamber at subsonic velocities and flowed through the first section of the chamber. As the gas entered the converging nozzle it accelerated and attained a velocity of Mach 1 at the throat. The gas continued to accelerate as it flowed through the diverging nozzle and reached velocities of approximately Mach 3.5. The gas then left the diverging nozzle and entered into a short (constant cross-sectional area) section, which was used to establish a constant gas velocity before impacting onto the target material. The gas then entered another expansion section and made contact with the target material. The gas flowed around the target material and exited out the chamber. A standing shock wave formed in the ex-

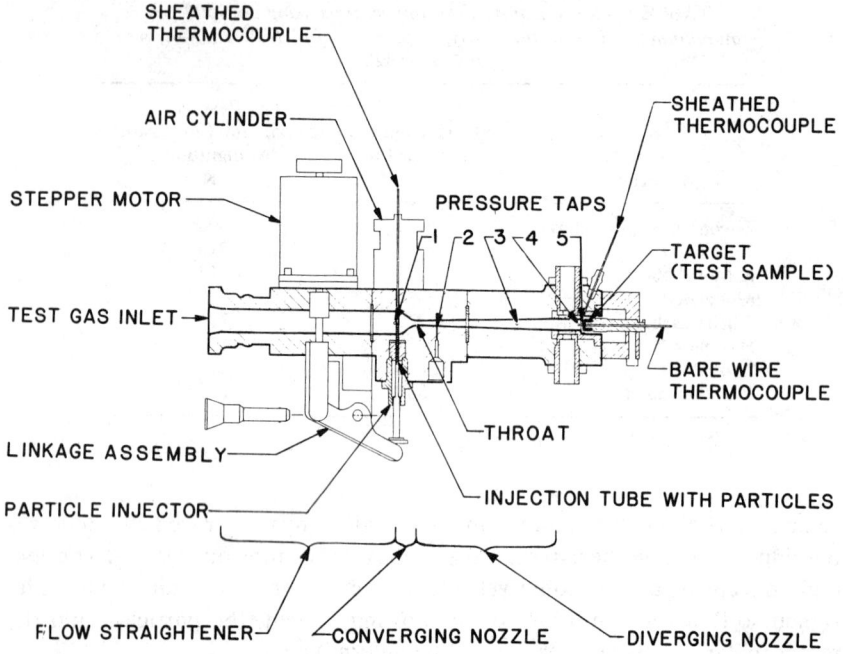

FIG. 2—*Schematic depicting the modified sonic velocity particle impact chamber.*

pansion portion a few millimetres upstream of the target material. A particle traveling through this shock wave had very little time to be affected by this disruptive flow region before impacting the target material.

Particles were injected via a mechanical injection system, which consisted of a stepper motor, linkage assembly, and particle injection tube, and a pneumatically operated retracting thermocouple. Particles were injected by simultaneously activating the stepper motor, which pushed the particles up the injection tube and retracted the thermocouple that held the particles in the injection tube. The injection process was controlled by a microprocessor. The particle injector tube was positioned in the gas stream directly in front of the converging nozzle centered on the nozzle throat.

The target material was placed on a mounting fixture directly in line with the gas flow stream. The target material was configured as a cup-like sample 0.95 cm (0.38 in.) in diameter and 0.16 cm (0.063 in.) thick and positioned in the chamber such that the flat surface was exposed to the impacting particles.

The oxygen pressure was measured using bonded strain gage pressure transducers at five locations in the chamber: entrance to the nozzle orifice, two locations in the diverging section of the nozzle, and two locations immediately before the target plate. The temperatures of the particles and the gas near the target material were measured using sheathed thermocouples, and the test target material temperature was measured using an exposed junction thermocouple. The velocity of the gas in the nozzle was estimated by a gas dynamic flow model developed for the chamber. The data from the instrumentation were digitally processed by a microprocessor and stored on a floppy disk. Data from each instrumentation channel were stored every 100 ms, which represents an averaged value obtained from eight data samples.

The particle impact chamber was interfaced to the NASA WSTF High Flow GOX Test System, which was capable of providing heated GOX at temperatures up to 755 K (900°F), at oxygen pressures up to 38 MPa (5500 psig), and maximum mass flow rates of 2.3 kg/s (5.1 lb/s).

Test Procedure

Particles were loaded into the injection tube and the target material positioned into the chamber. The test conditions were entered into the facility control microprocessor, which upon command activated the test facility and established the specified test conditions. Heated GOX at the test pressure was allowed to flow until the test temperature of the target material was achieved and the flow stabilized. Upon command, the microprocessor then activated the particle injection system and one particle or multiple particles were injected into the chamber. Oxygen flow was terminated 4 s after particle injection by the microprocessor, and instrumentation was checked for ignition of the target material. If no ignition occurred, the microprocessor automatically reactivated the facility and prepared the target material for another impact sequence. If an ignition occurred, the microprocessor safed the test data and a new target material was loaded into the chamber.

Test Results

Gas and Particle Velocity

Velocity of the oxygen was determined from a gas dynamics model developed specifically for the impact chamber. The accuracy of the model was checked (Table 6) by comparing the predicted pressure drops in the chamber with measured values at three point in the chamber, that is, designated as positions 1, 2, and 3 as shown in Fig. 2. The results show good agreement between the predicted and measured pressure values. Based on these results, the model is capable of predicting oxygen pressure or velocity within 10%.

The velocity of the particles was also predicted by this model. The predicted values were checked by conducting dent tests using copper and stainless steel particles. The results are given in Table 7 and indicate good agreement between the predicted and measured values. The 1600-μm aluminum particles were accelerated to approximately 50% of the gas velocity whereas the Type 304 stainless steel particles (1600 μm) were accelerated to approximately 35%

TABLE 6—*Comparison of predicted and measured pressures in impact chamber for inlet oxygen temperature of 550 K (530° F).*

Position 1[a]				Position 2[a]				Position 3[a]			
Pressure, MPa			Oxygen Velocity, m/s	Pressure, MPa			Oxygen Velocity, m/s	Pressure, MPa			Oxygen Velocity, m/s
Mod.	Exp.	(SD)		Mod.	Exp.	(SD)		Mod.	Exp.	(SD)	
34.4	35.1	(0.23)	10.8	18.1	17.1	(0.2)	410	0.77	0.85	(0.08)	807
27.6	27.6	(0.65)	10.8	14.5	13.7	(0.1)	410	0.67	0.69	(0.07)	823
20.7	21.1	(0.7)	10.9	10.9	10.6	(0.3)	410	0.52	0.51	(0.07)	870

[a]See Fig. 2 for locations of the three positions in the chamber where pressure was measured. Exp. = experimentally measured value. Mod. = predicted values from model. SD = standard deviation of measured values. Measurement errors for pressure transducer are Position 1: ±0.7, Position 2: ±0.7, and Position 3: ±0.04.

TABLE 7—*Comparison of predicted and measured particles.*

Impact Velocities, m/s	
Velocities Predicted by WSTF Model	Velocities Calculated by NWC Ballistic Equations Based on Depth of Penetration in Copper Dent Tests
277.4	304.8
261.2	281.9
251.5	277.4
216.4	259.1
216.4	251.5
216.4	237.7
181.4	205.7
181.4	181.4
167.6	181.4

of the gas velocities. The velocity of the 1600-μm aluminum and 304 stainless steel particles are given in Fig. 3 as a function of inlet pressure P_{in} and inlet temperature T_{in} for oxygen.

Oxygen Pressure

As oxygen was accelerated through the chamber, the pressure decreased drastically (Table 6). The oxygen pressure again increased in the region near the target material as the gas decelerated.

The pressure in this region, which will be referred to as target pressure P_t, is important because it was at this location where the ignition process takes place. Pressure was measured at two locations in this region, 0.10 and 0.32 cm (0.039 and 0.13 in.) from the target material and varies depending on the P_{in} and T_{in}. Figure 4 gives the relationship of P_t as function of P_{in} and the target material bulk temperature T_t. The value of P_t increased as P_{in} or T_t or both was increased.

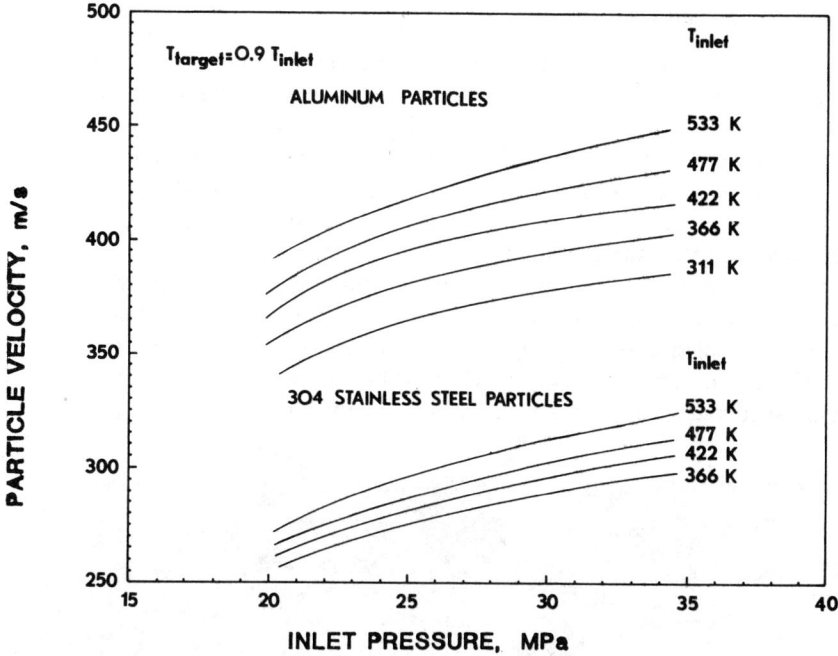

FIG. 3—*Particle velocity as a function of inlet gas pressure and temperature.*

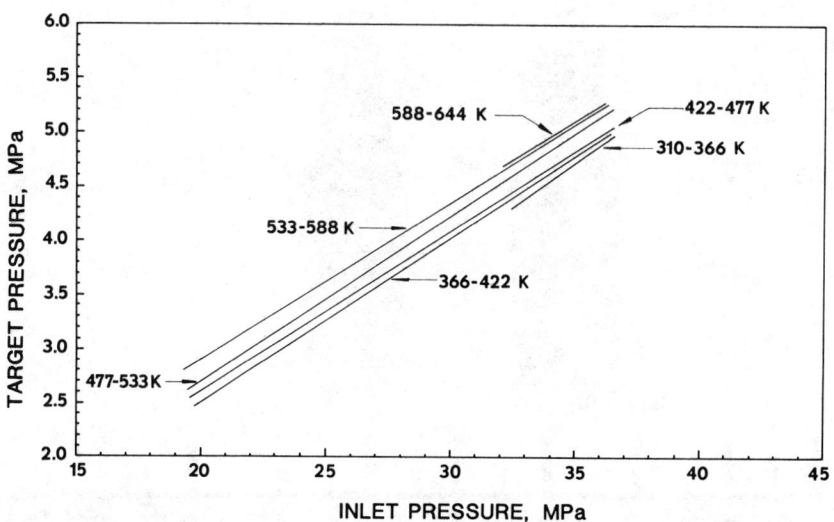

FIG. 4—*Target pressure as a function of inlet pressure and temperature.*

Definition of an Ignition

Ignition of a material in these tests was defined as an event that resulted in significant burning and consumption of the target material, rapid increase in temperature of the target material, and a visually observed fire at the exit of the chamber. In some cases, a brief flash was observed at the exit of the chamber, and post-test examination of the target material revealed only a slight amount of material was consumed at the target surface. This event was defined as a partial surface burn in which conditions (P_t and T_t) were not conducive for steady state combustion of the material. At present, this event is not being considered as a condition that would lead to a self-propagating fire. However, ongoing testing may provide data that will require a reexamination of this event. Typical target specimens that did not burn or only indicated partial surface burns are shown in Figs. 5a through e.

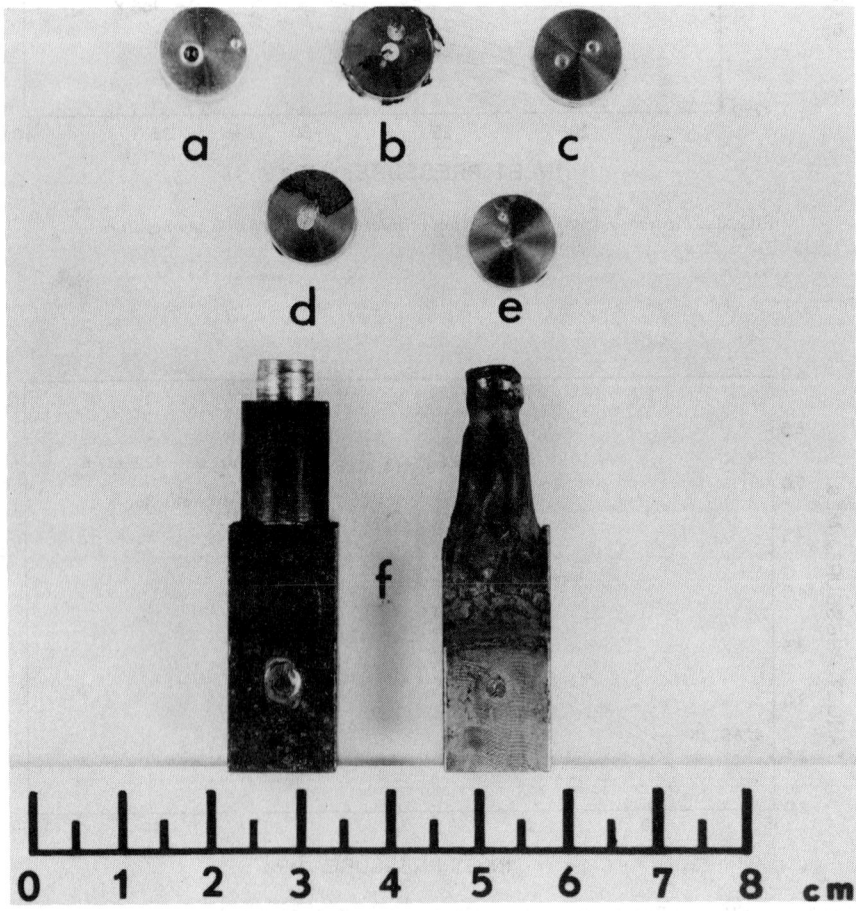

FIG. 5—*Photograph of typical post-test specimen.*

Figure 5a shows an aluminum 6061 specimen with an embedded 1600-μm stainless steel particle. Figure 5b shows a Monel 400 specimen that was impacted by two 1600-μm aluminum 2017 particles and illustrates the case of partial surface burning. Figure 5c shows a Type 316 stainless steel specimen that was impacted by two 1600-μm Type 304 stainless steel particles and illustrates the case of no ignition. Figure 5d shows an Inconel 718 specimen that was impacted by two 1600-μm aluminum 2017 particles in which the specimen exhibited partial surface burning. Figure 5e shows a Type 316 stainless steel specimen impacted by two 1600-μm aluminum 2017 particles in which the specimen exhibited no evidence of burning. Figure 5f shows the specimen holder and test specimen before and after a burn. In this case as with all tests that resulted in an ignition of the test specimen, the test specimen is entirely consumed.

Ignition Susceptibility of Material to Particle Impact

The ignition susceptibility (based on the specimen temperature required for ignition) of several metals and alloys at a nominal P_{in} of 27.5 MPa (4000 psig) is shown in Fig. 6. The results represent test data in which one spherical 1600-μm aluminum 2017 particle was impacted on the target material at various T_ts. Table 8 gives the ranking (in order of decreased susceptibility to ignition) of the materials in Fig. 6 compared with the results of similar materials from previous testing (see Background Section). Aluminum and the stainless steels were observed to ignite easily. Inconel 718 was much harder to ignite whereas Monel 400 did not ignite at T_ts up to 700 K (800°F). These same general observations were reported by other investigators [16,19,20].

Effects of Oxygen Pressure and Temperature

The effect of varying P_{in} and T_t on the ignition susceptibility of Type 316 stainless steel is illustrated in Fig. 7. As P_{in} was increased from approximately 21 to 35 MPa (3000 to 5100 psig), T_t required for ignition decreased from 585 to 360 K (583 to 188°F), respectively. The oxygen pressure in the region near the target material P_t is much less than P_{in} and is a function of P_{in} and T_{in} (Fig. 4). In reality, it is the oxygen pressure in this region that was involved in the ignition process. Figure 8 illustrates the ignition, and no ignition region for the results given in Fig. 7 in terms of P_t and T_t. The values of P_t were obtained from Fig. 4. The results in Fig. 8 indicate that ignitions actually occurred at pressures between approximately 2.9 and 5.0 MPa (421 and 726 psig), and as P_t was increased T_t required for ignition decreased. The implications that can be drawn from the data illustrated in Fig. 7 or 8 are discussed further in the Discussion of Results Section.

Effects of Particle Size and Type

Testing is presently being conducted to determine the effects of particle size and type on the ignition characteristics of material. The data presented repre-

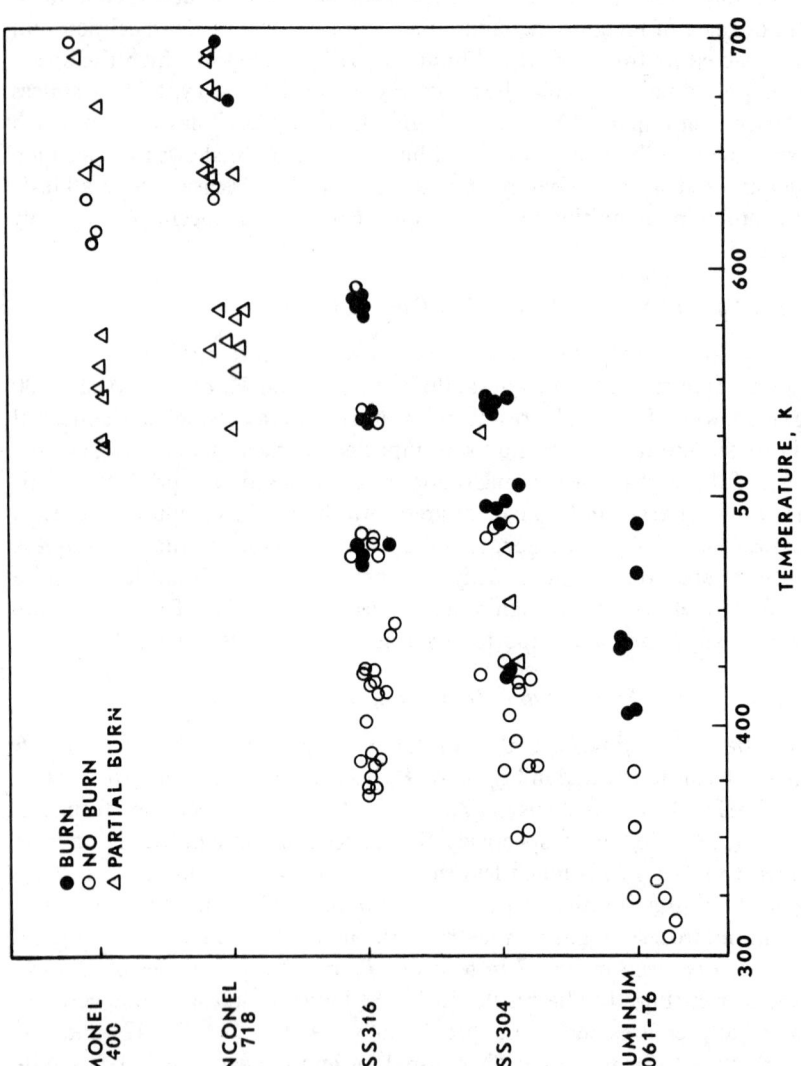

FIG. 6—Ignition and nonignition temperatures for several materials.

TABLE 8—*Ignition susceptibility of metals and alloys to particle impact.*

This Work	Ref 16[a]	Ref 19[b]	Ref 20[c]
Aluminum	aluminum	304 SS	316 SS
304 SS	stainless steel	Inconel 718	Monel (400, K-500)
316 SS	Monel alloys	Monel 400	. . .
Inconel 718	. . .	. . .	. . .
Monel 400	. . .	. . .	. . .

[a]Ref 16, condition not specified.
[b]Ref 19, multiple 800-μm aluminum particles, sonic gas velocity.
[c]Ref 20, ten 1600-μm aluminum particles, sonic gas velocity.

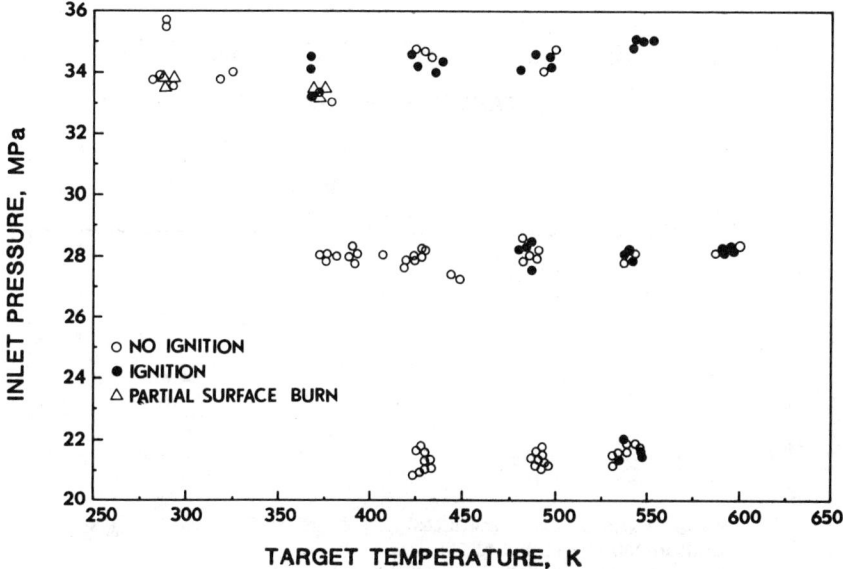

FIG. 7—*Effect of pressure on the ignition of Type 316 stainless steel.*

sent only a portion of the test matrix that is planned. To date, testing has been conducted in which spherical 1600- and 2000-μm (0.063- and 0.079-in.) diameter aluminum 2017 particles were impacted on Type 316 stainless steel at P_{in}s between 27 and 29 MPa (3900 and 4200 psig). The results are shown in a plot of P_{in} versus T_t (Fig. 9). The T_ts required for ignition of Type 316 stainless steel with the 1600-μm aluminum particles were greater than 480 K (404°F). Below this temperature, no ignitions out of ten impacts were observed.

In case of impacting the 2000-μm aluminum particles on Type 316 stainless steel, 3 ignitions out of 19 impacts were observed below 480 K. The lowest T_t observed to date for ignition with the 2000-μm particles was approximately

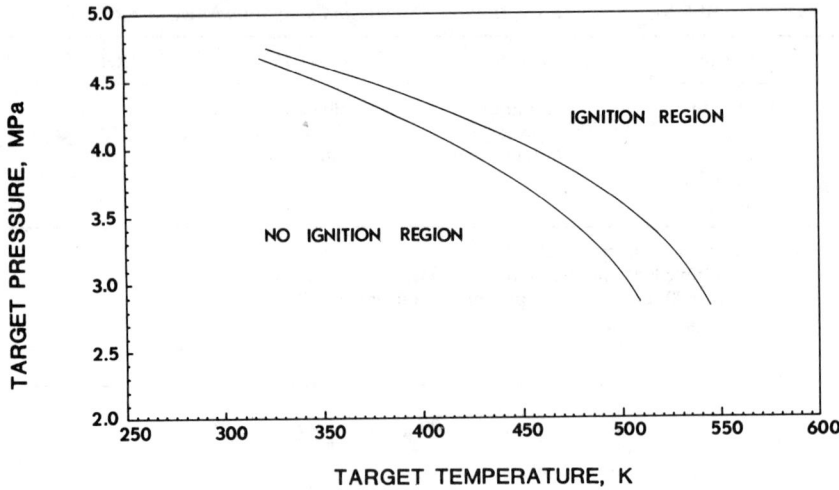

FIG. 8—*Effect of target oxygen pressure on ignition of Type 316 stainless steel.*

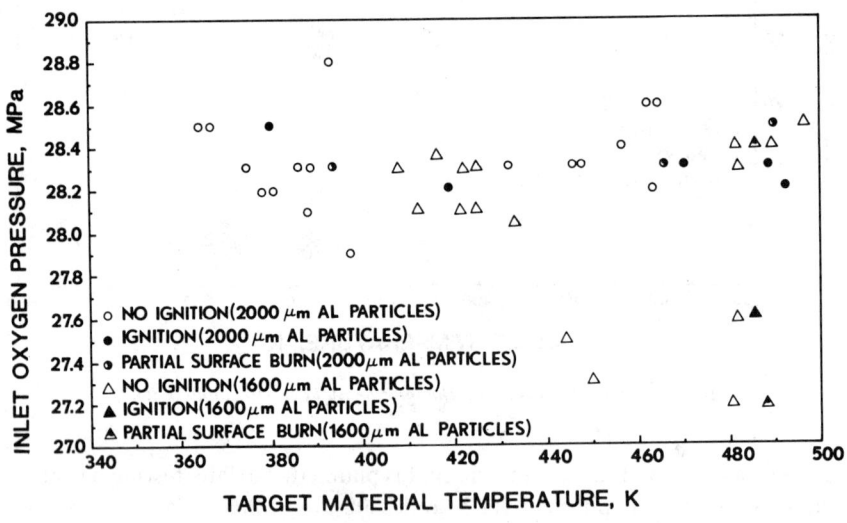

FIG. 9—*Effect of particle size of the ignition of Type 316 stainless steel.*

380 K (224°F). In addition, 2 out of the 16 impacts (2000-μm particles) that did not produce ignitions resulted in a slight surface burning. These results strongly suggest that the larger particles ignited the Type 316 stainless steel more easily than the 1600-μm particles. These results also agree with results from previous particle impact tests (see Background Section Ref *19*), which indicated that 800-μm (0.031-in.) diameter aluminum particles ignited mate-

rials more easily than 150-μm (0.0059-in.) diameter aluminum particles. However, this trend may change with very small particles (less than 100-μm [0.0039-in.] diameter particles) based on previous testing reported by Nihart and Smith [16] and Wegener [4]. More testing is required before a definitive conclusion on the effects of particle size can be made.

A limited number of tests have been conducted using spherical 1600-μm Type 304 stainless steel particles. Figure 10 shows the results of the impact tests using 1600-μm Type 304 stainless steel and aluminum 2017 particles at P_{in}s between 27 and 29 MPa (3900 and 4200 psig). No ignitions were observed using the Type 304 stainless steel particles at T_ts that produced ignitions using the aluminum particles. Impact tests conducted on aluminum 6061 using 1600-μm Type 304 stainless steel and aluminum 2017 particles are shown in Fig. 11. The results indicated that higher T_ts were required for ignition using the stainless steel particles then the aluminum particles.

Discussion of Results

One of the objectives of the present test program at WSTF is to determine the major factors involved in the ignition process of materials by particle impact. Impact testing to date has revealed some important trends that will help this determination in the coming years. Before an ignition mechanism can be determined for particle impact type ignitions, the energetics (energy balance)

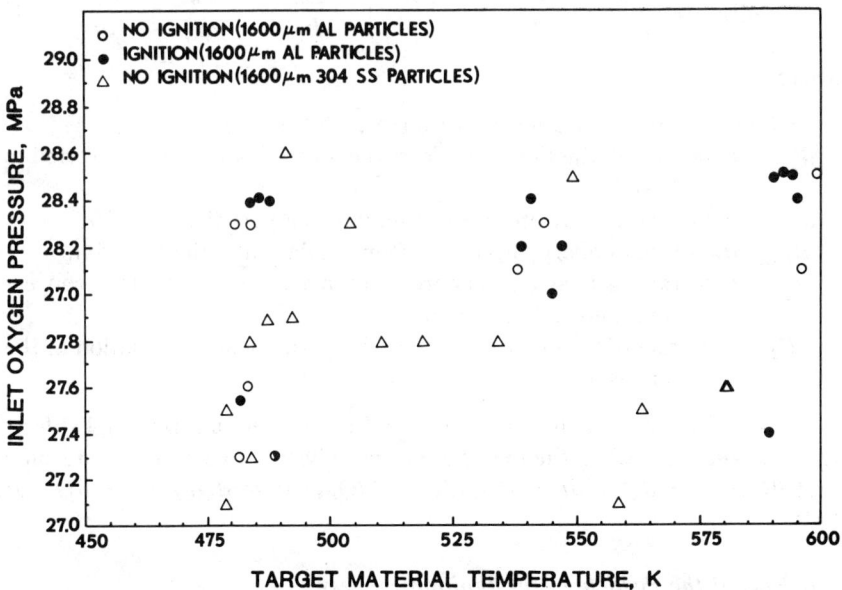

FIG. 10—*Effect of particle type on the ignition of Type 316 stainless steel.*

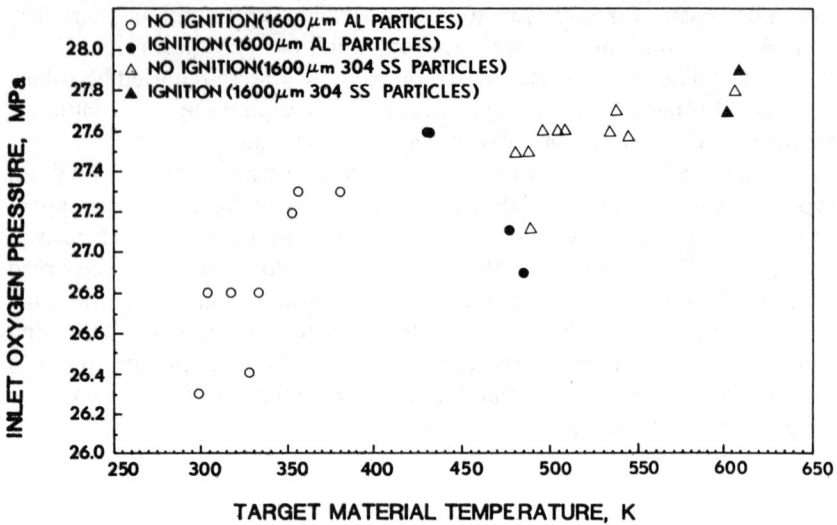

FIG. 11—*Effect of particle type on the ignition of aluminum 6061.*

of the ignition process must be defined. The rate of heat that will accumulate in a material (dQ_{acc}/dt) from a particle impact process is illustrated by Eq 1.

$$dQ_{acc} = \frac{dQ_E}{dt} + \frac{dQ_{KP}}{dt} + \frac{dQ_{HP}}{dt} + \frac{dQ_{RX}}{dt} - \frac{dQ_{HT}}{dt} - \frac{dQ_w}{dt} \quad (1)$$

where

dQ_E/dt = rate of external heat sources,
dQ_{KP}/dt = rate of kinetic energy from the impacting particle converted to heat,
dQ_{HP}/dt = rate of heat produced from burning particles,
dQ_{RX}/dt = rate of energy produced from oxidation of the material,
dQ_{HT}/dt = rate of heat transfer from the material (conduction, convection, and radiation), and
dQ_w/dt = rate of heat converted to work (permanent deformation of the material).

Factors that affect each of the terms in Eq 1 are summarized in Table 9. Ignition will occur when the material has been heated to a temperature such that $dQ_E/dt + dQ_{KP}/dt + dQ_{HP}/dt + dQ_{RX}/dt = dQ_{HT}/dt + dQ_w/dt$ [21].

The Role of the Particle in the Ignition Process

When a particle (metal or alloy) impacts on a test specimen, two energy sources are available to heat the specimen to its ignition temperature: kinetic

TABLE 9—*Factors affecting particle impact type ignition.*

Factor	Definition
$\dfrac{dQ_E}{dt}$	1. Temperature and thermal capacity of the oxygen 2. Heat transfer coefficient between gas and target surface
$\dfrac{dQ_{KP}}{dt}$	1. Velocity of particle 2. Mass of the particle 3. Type of collision (elastic or inelastic) 4. Impact obliquity
$\dfrac{dQ_{HP}}{dt}$	1. Heat of combustion of the particle 2. Mass of particle 3. Kinetics of the particle burning process 4. Fraction of energy imparted to the target material
$\dfrac{dQ_{RX}}{dt}$	1. Kinetics of the oxidation process prior to the ignition phase 2. Heats of combustion of the target material
$\dfrac{dQ_{HT}}{dt}$	1. Conduction heat loss from impact surface 2. Convection heat loss from impact surface 3. Radiation heat loss from impact surface
$\dfrac{dQ_w}{dt}$	1. Mechanical properties of the target material and particle 2. Type of collision

energy of the particle Q_{KP} and energy released from combustion of the particle Q_{HP}. The magnitudes in energy for Q_{KP} and Q_{HP} are illustrated in Table 10. Note that the Q_{HP}s are over two orders of magnitude greater than the Q_{KP}s. In the case of impacting aluminum particles, burning particles were observed exiting the chamber after impacts that did not result in ignition of the target material. It is therefore believed that the aluminum particles also ignited in the impact tests that resulted in ignition of the target material. There was no evidence that the Type 304 stainless steel particles ignited. If ignition of a material requires the heat that is produced by the burning of the impacting particle, then the impact of an aluminum particle should ignite materials more easily than a 304 stainless steel particle. The results in Fig. 10 illustrate that aluminum particles definitely ignite Type 316 stainless steel more easily than Type 304 stainless particles. In fact ignition of Type 316 stainless steel could not be induced with Type 304 stainless steel particles at T_is 100 K (180°F) greater than the lowest T_t required for ignition using alu-

TABLE 10—*Summary of particle impact energies for aluminum and 304 stainless steel particles.*

Particle	Kinetic Energy, J[a]	Heat of Combustion, J
1600-μm aluminum	0.49	176.0
2000-μm aluminum	0.85	342.0
1600-μm stainless steel	0.75	90.0

[a]Velocities of the particles were determined at $P_{in} = 27.6$ MPa and $T_{in} = 478$ K.

minum particles. Similar results were observed by Porter [*19*] in which ignition of materials was obtained with 800-μm (0.32-in.) aluminum particles but not with similar sized Inconel 718 particles.

One could argue that the collision of a Type 304 stainless steel particle converts less of its kinetic energy to heat than aluminum particles because of differences in their collision properties. However, in a test in which a Type 304 stainless steel particle was impacted on aluminum 6061, the particle implanted itself into the aluminum target without causing ignition of itself or the aluminum target (Fig. 5*a*). In this case, it can be assumed that the majority of the kinetic energy of the particle was converted to heat and sample deformation. Similar tests were conducted using aluminum particles, and the results (Fig. 11) show that aluminum particles caused the aluminum 6061 targets to ignite at T_ts 125 K (225°F) lower than the T_ts required for ignition using the stainless steel particle. These results suggest that in the case of impacting Type 304 stainless steel particles on Type 316 stainless steel targets (Fig. 8), ignitions of the stainless steel targets were not obtained because the particles contained insufficient kinetic energy to ignite and burn.

As shown in Fig. 11 aluminum 6061 was ignited with the stainless steel particles; however, the ignition mechanism is not fully understood. To date, there is no evidence that the stainless steel particles ignited and burned. While it is likely that the bulk of the kinetic energy of the particle is imparted to the specimen, the ignition could be from that energy alone. However, during the impact process, a crater was formed on the aluminum target that exposed fresh (unoxidize) metal. Monroe et al [*18*] demonstrated that ignition of aluminum can be enhanced by the sudden exposure of fresh metal to oxygen.

Based on the results to date, it is believed that the kinetic energy of the particle converted to heat is utilized to ignite the particle. Once the particle ignites, the particle must reside at or near the surface of the material so that heat can be transferred to material. Thus the energy flux density during the impact and subsequent particle burning is a critical parameter for describing the ignition mechanism. However, materials, such as aluminum impacted with hard to ignite particles, may follow a different mechanism.

The Role of Oxygen Pressure on the Ignition Process

As the inlet oxygen pressure P_{in} is increased in the supersonic impact chamber at a fixed inlet temperature T_{in}, the mass flow rate, gas and particle velocities, and pressure near the target P_t will increase. The effects of increasing P_{in} on the ignition process are (1) increased kinetic energy of the particle, (2) increased preignition oxidation, (3) increased burning rate of the particle and target material, and (4) increased convective heat loss. Items 1 and 3 would increase the susceptibility of ignition for a material whereas Item 2 may increase or decrease ignition susceptibility depending on whether nonprotective or protective oxide coatings form on the surfaces of the material and particles. Item 4 will decrease the ignition susceptibility of a material.

The results that were given in Fig. 7 or 8 indicate that as P_{in} or P_t was increased, the T_t required for ignition of Type 316 stainless steel decreased. One could assume from these data that increasing P_{in} lowered the ignition temperature threshold for Type 316 stainless steel. This assumption is consistent with a decrease in the ignition temperature of stainless steel as oxygen pressure is increased [6]. These data also imply that the increase in pressure increases the oxidation rate (dQ_{RX}/dt). Thus, for a given heat loss rate (dQ_{HT}/dt), a lower initial T_t is required to generate a sufficient heat (dQ_{acc}/dt) necessary for ignition of the target material.

The T_t was varied by varying the T_{in} of oxygen, and because of the supersonic conditions in the chamber, the particle velocity varied depending on P_{in} and T_{in} of oxygen (Fig. 3). Thus there exists different combinations of P_{in} and T_{in} in which the gas velocity will be the same. In the previous section, it was shown that the energy released from a burning particle was required to ignite the target material and that the kinetic energy of the particle was utilized to ignite the particle. Therefore the decrease in T_t as P_{in} was increased (Fig. 7 or 8) may in reality reflect the kinetic energy level required to ignite the particle and not the target material. If this supposition is correct, then the particle velocity for the ignition and nonignition regions in Fig. 7 or 8 would be essentially the same.

To prove or disprove this supposition, the particle velocities were determined along the region defining the ignition and nonignition of Type 316 stainless steel (Fig. 7) by using velocities given in Fig. 3 and the relationship: $T_t = 0.9\ T_{in}$. The results indicated that the particle velocities along this ignition curve were 409 $\pm$ 12.4 m/s (1341 $\pm$ 41 ft/s) or equivalent to kinetic energies of 0.49 $\pm$ 0.0004 J. These results suggest that as the P_{in} of oxygen was decreased, the T_t, or in reality the T_{in}, of oxygen required for ignition increased to compensate for the decrease in kinetic energy of the particle because of the decrease in P_{in}. However, additional data are required before a definitive conclusion can be drawn.

Conclusions

Ignition of metals and alloys have been obtained by the impact of high-velocity particles in a supersonic oxygen flow stream. Aluminum 6061, Type 316 stainless steel, Type 304 stainless steel, and Inconel 718 were ignited with 1600-μm aluminum 2017 particles at elevated temperatures and pressures whereas Monel 400 could not be ignited. The ignition susceptibility of metals and alloys appeared to increase as the particle size was increased. Type 304 stainless steel particles required more extreme conditions for ignition of materials than similar sized aluminum particles. The effects of increasing oxygen pressure appeared to decrease the required target specimen temperature necessary for ignition. However, because of the supersonic conditions in the chamber, the decrease in temperature may be a result of the changes in the gas velocity or kinetic energy of particles. The results indicated that ignition and subsequent burning of the particles were required for ignition of the target materials with the possible exception of aluminum 6061. The energy flux density during impact and subsequent burning of the particles appears to be one of the major factors governing the ignition process.

Future Testing

Testing will continue in the future to determine the effects of particle size (less than 1000 μm), type, and geometry on the ignition of materials. The effects of the collision process (elastic or inelastic) on ignition of particles and target materials will also be evaluated. The impact chamber will be modified to investigate the ignition characteristic of materials at subsonic gas velocities. The effects of material surface alterations on ignition by particle impact will be studied. As testing progresses, a model of the ignition process will be developed.

Acknowledgments

The authors wish to acknowledge Jill Rollings and Charles Murray who conducted the majority of the tests from which the data presented in this paper were obtained. In addition, the authors would like to extend their appreciation to Joel Stoltzfus and Dr. Michael Pedley for their many hours of consultation. Special acknowledgment is extended to Bill Porter for his efforts in developing the flow dynamics model from which the chamber was designed and evaluated. The funds for this work have been provided by the Office of the Chief Engineer, NASA Headquarters, Washington, DC, primarily through the efforts of Joyce McDevitt.

References

[1] Lapin, A., *Liquid and Gaseous Oxygen Safety Review*, Vol. I-IV, NASA-CR-120922, APCI TM184, NASA Lewis Research Center, Cleveland, OH, June 1972.

[2] Clark, A. F. and Hurst, T. G., "A Review of the Compatibility of Structural Materials with Oxygen," *AIAA Journal*, Vol. 12, No. 4, April 1974, pp. 441-454.

[3] Oxygen Fire Event Report File, Laboratory Test Office, NASA White Sands Test Facility, Las Cruces, NM.

[4] Wegener, W., "Investigations on the Safe Flow Velocity to be Admitted for Oxygen in Steel Pipe Lines," *Stahl and Eisen*, Vol. 84, No. 8, 1964, pp. 469-475.

[5] Yuen, H. H. Compatibility of Materials with Oxygen NAEC Report: NAEC MISC 92-0354, Naval Air Engineering Center, Lakehurst, NJ, Sept. 1978.

[6] Schmidt, H. W. and Forney, D. F., "ASRDI Oxygen Technology Survey, Volume IX: Oxygen Systems Engineering Review," NASA SP-3090, NASA, Washington, DC, 1975.

[7] Williams, R. E., The Combustion of Laser Ignited Zirconium, final report NSG 518, NASA, Washington, DC, 1967.

[8] Williams, R. E. and Kottenstette, J., *Atomic and Molecular Absorption Spectroscopy in the Combustion of Magnesium Particles*, Fifth Mid-America Symposium on Spectroscopy, SAS, 1967.

[9] Williams, R. E., McLain, W., and Walter, L., "The Ignition and Combustion of Single Boron Particles in Air at Ambient Pressure," CPIA 204, Vol. 1, Seventh JANNAF Combustion Meeting, San Diego, CA, 1971.

[10] Williams, R. E., Kelley, C., and Takemoto, A., "The Combustion Kinetics of Particulate Boron," final report, Project SQUID, Office of Naval Research, 1971.

[11] Wolfhard H. G., Glassman, I., and Green, L., *Heterogeneous Combustion*, Academic Press, New York, Washington, DC, 1964.

[12] Prentice, J. L., Drew, C. M., and Christensen, J. "Preliminary Studies of High Speed Photography of Aluminum Particles Combustion Flame," *Pyrodynamics*, Vol. 3, 1965, pp. 81-90.

[13] Andersen H. and Belz, L. H., "Factor Controlling the Combustion of Zirconium Powders." *Journal of Electrochemical Society*, Vol. 10, No. 5, May 1953, pp. 240-249.

[14] Bartlett, R. W., Ong, J. W., Fassell, W. M., and Papp, C. A., "Estimating Aluminum Particle Combustion Kinetics," *Combustion and Flame*, Vol. 7, 1963 pp. 227-234.

[15] Law, C. K., "Models for Metal Particle Combustion with Extended Flame Zones," *Combustion Science and Technology*, Vol. 12, 1976, pp. 113-124.

[16] Nihart, G. J. and Smith, C. P., "Compatibility of Metals with 7500 psi Oxygen," Report AD 608 260, Union Carbide, Linde Division, Tonawanda, NY, Oct. 1964.

[17] Phillips, B. R., "Resonance Tube Ignition of Metals," Ph.D. thesis, Department of Engineering, University of Toledo, July 1975, p. 198.

[18] Monroe, R. W., Bates, E., and Pears, C. D., "Structural Materials Evaluation for Oxygen Centrifugal Compressors," final report SORI-EAS-78-339, Southern Research Institute, Birmingham, AL, Sept. 1978.

[19] Porter, W. S., "Metals Ignition Study in Gaseous Oxygen (Particle Impact Technique Relating to the Shuttle Main Propulsion System Oxygen Flow Control Valve)," NASA WSTF Report TR-277-001, NASA White Sands Test Facility, Las Cruces, NM, Oct. 1982.

[20] Schoenman, L. "Selection of Burn-Resistant Material for Oxygen-Driven Turbopumps," paper presented at the 20th AIAA/ASME/SAE Joint Propulsion Conf., Cincinnati, OH, 1984.

[21] Glassman, I., Mellor, A. M., Sullivan, H. F., and Laurendeau, N. M., "A Review of Metal Ignition and Flame Models," Conference Proceedings 52, NASA AGARD Annual Meeting, Neuilly-Sur-Seine, France, Feb. 1970, pp. 19 (1-30).

Frank J. Benz[1] *and Joel M. Stoltzfus*[2]

Ignition of Metals and Alloys in Gaseous Oxygen by Frictional Heating

REFERENCE: Benz, F. J. and Stoltzfus, J. M., **"Ignition of Metals and Alloys in Gaseous Oxygen By Frictional Heating,"** *Flammability and Sensitivity of Materials in Oxygen-Enriched Atmospheres: Second Volume, ASTM STP 910*, M. A. Benning, Ed., American Society for Testing and Materials, Philadelphia, 1986, pp. 38–58.

ABSTRACT: The ignition of metals and alloys have been investigated by rotating the end of a hollow cylinder against an identical stationery cylinder on a common axis (frictional heating) in gaseous oxygen. A ranking criterion that measures the resistance of metals and alloys to ignition is discussed. It consists of the power per unit area required for ignition, conveniently expressed as the product of contact pressure P and average linear velocity $\bar{v}$. Data are presented that demonstrate that materials that are high in nickel and copper require greater Pv products for ignition (more resistant to ignition) than materials that are high in iron. Aluminum and titanium alloys are shown to require the lowest Pv products for ignition. The effects of varying surface velocity, contact pressure, coefficient of friction, and oxygen pressure on $P\bar{v}$ products required for ignition are discussed. The results indicate that $P\bar{v}$ product required for ignition increases as surface velocity increases, as contact pressure decreases, and as the coefficient of friction decreases. Increasing oxygen pressure will cause the $P\bar{v}$ product required for ignition to decrease at low pressures and increase at high pressures. In some cases, the relative ranking of materials based on $P\bar{v}$ products will also change as conditions are varied.

The data presented in this paper are from a development program to determine the merits of using such a dynamic test method for ranking materials for oxygen service. The results to date have indicated that more testing is required before this method can be made into a standard test. The data presented do show some of the major parameters affecting the ignition of materials when exposed to frictional heating. The repeatability of the data at any set of conditions have not been thoroughly investigated.

KEY WORDS: ranking, carbon steels, stainless steels, copper alloys, nickel alloys, brasses, aluminum alloys, titanium alloys, friction ignition, oxygen compatibility, ignition criteria

[1]Materials engineer, NASA/JSC White Sands Test Facility, P.O. Drawer MM, Las Cruces, NM 88004.
[2]Engineer, Lockheed-EMSCO, White Sands Test Facility, P.O. Drawer MM, Las Cruces, NM 88004.

Nomenclature

A — Cross-sectional area, M^2
C — Heat-capacity, J/kg K
$C(T)$ — Heat capacity as a function of temperature, J/kg K
L — Torque, Nm
P — Contact pressure, N/m^2
Pox — Oxygen pressure, MPa
$\dot{Q}_{fric}$ — Rate of heat caused by friction, J/s
$\dot{Q}_L$ — Rate of heat loss, J/s
$\dot{Q}_{RX}$ — Rate of heat caused by oxidation, J/s
$\dot{Q}w$ — Rate of energy loss caused by work; permanent sample deformation, J/s
T — Temperature, K
Tb — Temperature at steady burning, K
Tc — Critical temperature, K
T_f — Temperature at which flame develops, K
T_{ig} — Ignition temperature, K
T_t — Transition temperature, K
i — Initial conditions
m — Mass, kg
n — Conditions at an arbitrary point
q — Heat flow, J/s
t — Time, s
dq_{acc}/dt — Rate of heat accumulation, J/s
$\bar{v}$ — Average linear surface velocity, m/s
μ — Coefficient of friction
ω — Angular velocity, rad/s

Introduction

The ignition of metals in oxygen caused by frictional heating has for many years plagued the designers and users of pumps, compressors, and various other oxygen components [1–7]. Only recently have attempts been made to develop test methods that evaluate the ignition of metals and alloys by frictional heating [3,8,9]. The National Aeronautics and Space Administration (NASA) White Sands Test Facility (WSTF) is presently involved in developing several test methods for evaluating the ignition and combustion process of materials in oxygen [10]. This paper will describe a friction rubbing apparatus that has been developed and will present the results that have been obtained to date.

Background

The energetics involved in a rubbing process of a material in an oxygen atmosphere are described by Eq 1. The rate of heat that accumulates in the material is a function of the rate of frictional energy applied to the material, rate of heat produced from oxidation, rate of heat loss from the material (conduction, convection, and radiation), and rate of energy loss caused by work (permanent deformation of the material)

$$dq_{acc}/dt = \dot{Q}_{RX} + \dot{Q}_{fric} - \dot{Q}_L - \dot{Q}_W \tag{1}$$

The temperature increase of the material at any time T_n can be determined by integrating Eq 1 where $dq_{acc} = mC(T)dT$ in Eq 2

$$T_n - Ti = \frac{\displaystyle\int_{ti}^{t_n}(\dot{Q}_{RX} + \dot{Q}_{fric} - \dot{Q}_L - \dot{Q}_W)dt}{\displaystyle\int_{Ti}^{T_n} mC(T)dT} \tag{2}$$

Ignition of the material is possible when sufficient heat has accumulated in the material to raise the temperature of the material such that $\dot{Q}_{RX} = \dot{Q}_L + \dot{Q}_W$. Glassman et al [11] defined this temperature as the critical temperature T_c. Ignition actually occurs when a flame appears T_f and steady state burning Tb develops.

Figure 1 illustrates the relationship of T_c and T_f in a plot of heat flow q versus surface temperature T_s for ignition of a material that produces gaseous products. Glassman et al suggested that the true ignition temperature T_{ig} occurs between T_c and T_f when the oxidation heat curve is parallel to the heat loss curve. Including the mechanical work term as well as the heat loss, this gives $d\dot{Q}_{RX}/dT = d\dot{Q}_L/dT + d\dot{Q}_W/dT$.

In the case of metals that form solid products, the oxidation rate can be inhibited by the formation of protective oxide coatings. The oxidation process of metals is complex, and entire texts have been written on the subject [12–14]. For this discussion, only three types of oxidation rates will be considered: rates dependent on the absorption of gaseous oxygen at the metal or oxide surface (linear oxidation rate law), rates dependent on electric-field induced transport of metal ions through n-conducting oxide coatings (parabolic oxidation rate law), and rates dependent on electric-field induced transport of metal ions through p-conducting oxide coatings (cubic oxidation rate law). Figure 2 illustrates the effects of protective oxide coatings on the ignition process. A new term must be introduced that reflects the transition temperature T_t where the protective coating fails (cracks, spalls, melts, or sublimes) and gives way to the faster linear oxidation rate. Figure 2 illustrates that for protective oxide coatings that fail at temperature below $T_c(T_t < T_c)$, the linear

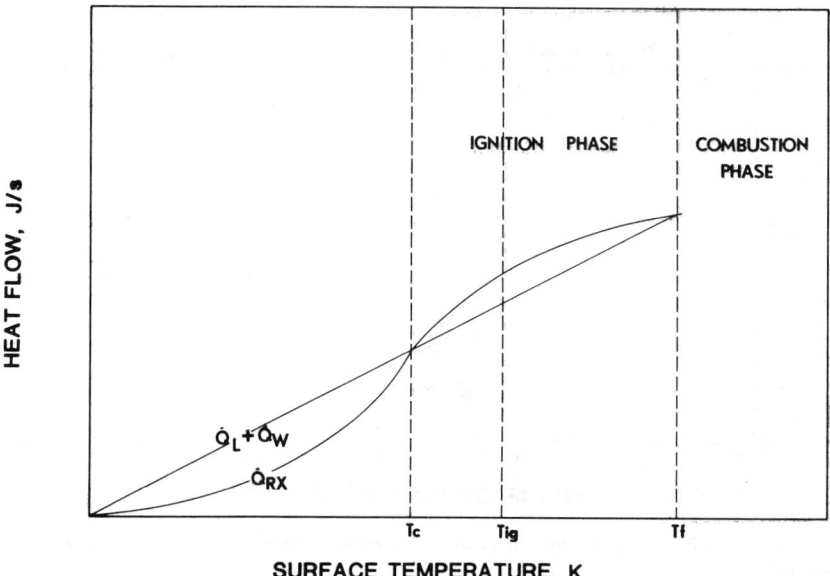

FIG. 1—*Ignition diagram for materials that form gaseous products* [11].

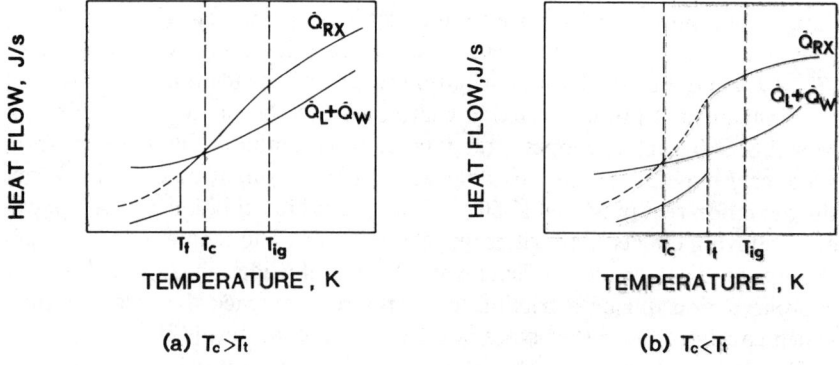

(a) $T_c > T_f$ (b) $T_c < T_f$

FIG. 2—*Ignition diagrams for materials that form solid products* [11].

oxidation rate will proceed to T_c and subsequent ignition. However, for metal oxides where $T_f > T_c$, failure of the oxide coating leads abruptly to ignition (Fig. 2b).

Figure 3 illustrates the results of Jenny and Wyssmann [9] for friction induced ignition of cast iron rubbed against Type 403 stainless steel and Monel K-500 rubbed against Monel K-500. Temperature measurements were limited to temperatures above 975 K (1400°F). These investigators concluded (at least for the measured temperature range) that the oxidation heat rate for the

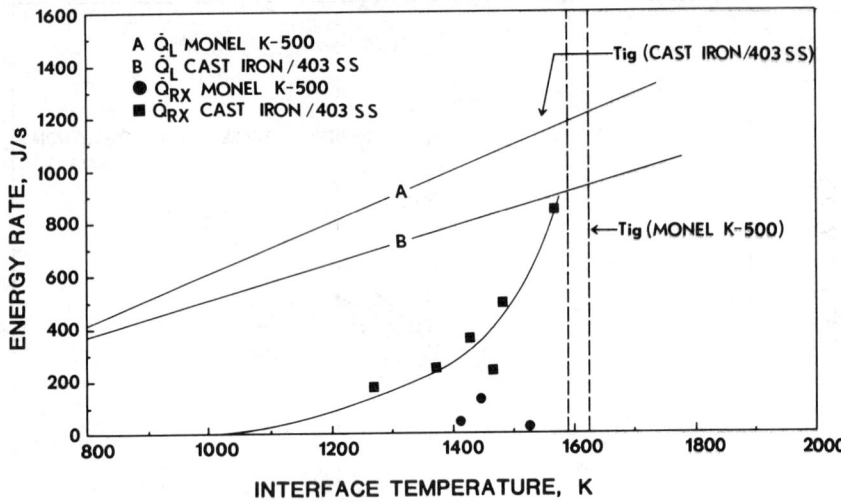

FIG. 3—*Ignition diagram for frictionally induced ignition of Monel K-500 and cast iron/403SS* [9].

cast iron/stainless steel was dependent on gaseous oxygen adsorption at the surface or the linear oxidation rate law. This is evident from their heat rate data before ignition, which are similar to Fig. 1 or Fig. 2a (where $T_t < T_c$). The heat rate data for the Monel K-500 before ignition resembles Fig. 2b ($T_t > T_c$) and implies that the oxidation rate was dependent upon a diffusion mechanism in a protective oxide coating. Since the major constituents of Monel K-500 (nickel, copper, manganese, and aluminum) in their pure form produce p-type conducting oxide coatings [13], it might be concluded that the oxidation rate of Monel K-500 follows the cubic oxidation rate law. However, alloying effects, impurities, oxygen pressure, and temperature will vary the type of diffusion mechanism involved in metal oxidation [12,13]. Also, the rubbing action during a friction test can stress or crack the oxide coating, which can introduce other types of diffusion mechanisms [12].

The ranking criteria used by Jenny and Wyssmann consisted of comparing the axial loads as a function of time required for ignition, as shown in Fig. 4. Monel K-500 required greater axial loads for ignition than the cast iron/steel and indicates that of the two materials, Monel K-500 is more resistant to ignition.

Many questions remain as to the major parameters that affect the ignition of metals by friction heating. Subjects that will be addressed in this paper are listed below:

(1) Development of a ranking criteria
(2) Effects of surface velocity and contact pressure
(3) Conditions affecting coefficient of friction

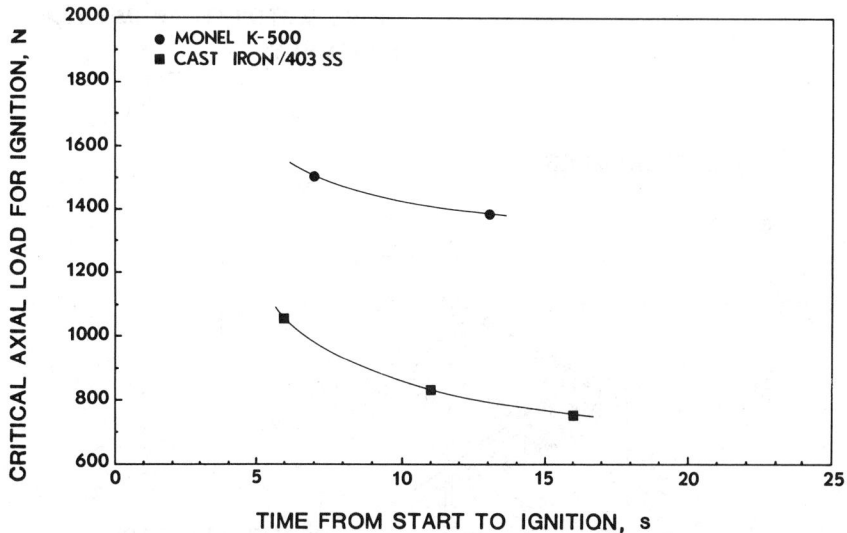

FIG. 4—*Axial load requirement for ignition of Monel K-500 and cast iron/403SS* [9].

(4) Effects of surface treatment
(5) The importance of heats of combustion
(6) Effects of oxygen pressure
(7) Effects of the rubbing process on ignition

Test System Description

The friction rubbing test system (Fig. 5) consists of a cylindrical pressure chamber (Fig. 6) fabricated from Monel 400, which contains an inner cavity approximately 3.8 cm (1.5 in.) in diameter and 6.4 cm (2.5 in.) in length. The chamber is provided with a rotating shaft that extends through the chamber via a series of bearings and seals. The shaft is connected at one end to a drive motor/transmission assembly which is capable of turning the shaft at rotational speeds over a range from 1000 to 20 000 rpm. The other end of the shaft is connected to an air actuator, which allows axial movement of the shaft and provides the capability of applying up to 4450 N (1000 lbf) load on the test samples. The test samples consist of two identical hollow cylinders, 2.5 cm (1 in.) outside diameter and 2.0 cm (0.8 in.) inside diameter. The rubbing surface area is approximately 1.8 cm² (0.28 in.²). The test samples are prepared by machining stock material to the specified dimensions. No sample surface finished was required for these tests since after the first few seconds of rubbing the surface is rubbed away. One sample is mounted to the rotating shaft and the other is affixed to a sample housing. The sample housing is attached to the inside of the chamber on bearings. Movement of the

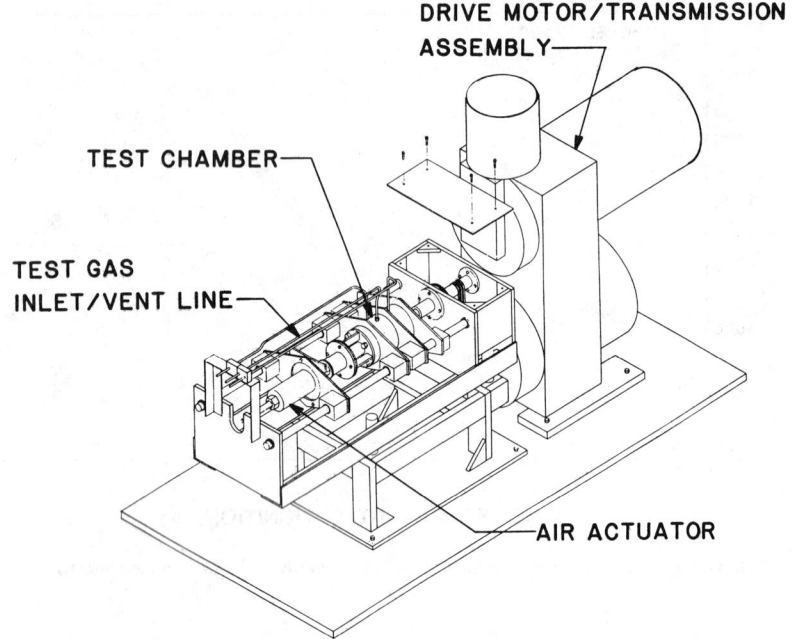

FIG. 5—*Friction rubbing apparatus.*

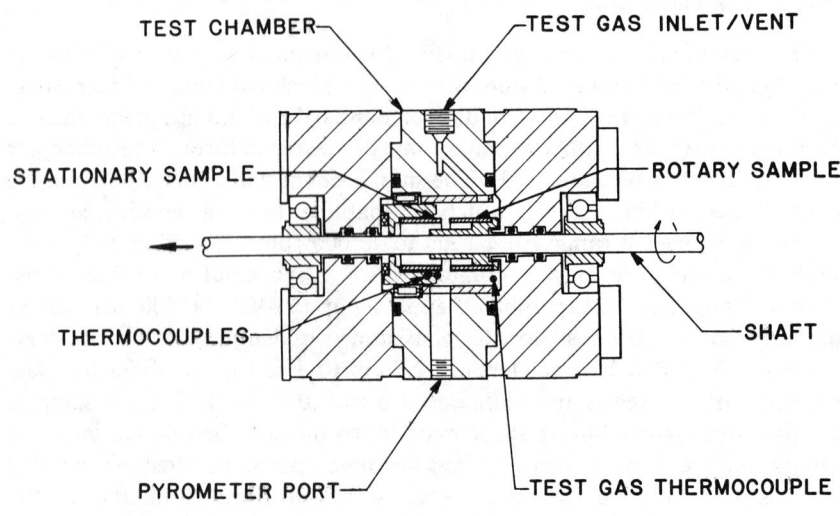

FIG. 6—*Cross-sectional view of the friction rubbing test chamber.*

sample housing is restrained via a pin positioned against a load cell. This design allows for direct measurement of the frictional force generated during the rubbing process. The test samples have a surface rubbing velocity of approximately 24.4 m/s (80 ft/s) at the maximum rotating speed of 20 000 rpm. The test gas is provided to the chamber via a 68.9-MPa (10 000-psig) gas distribution system.

Pressure in the chamber and pressure to the air actuator are measured using bonded stain gage pressure transducers. Temperature of the oxygen and the fixed test sample are measured using sheathed Chromel-Alumel thermocouples. The temperatures of the rubbing surfaces above 1200 K (1700°F) are measured using a two-color optical pyrometer. A thermopile is used to detect changes in heat radiation from the rubbing samples and to verify the point of ignition. Sample load and sample torque are measured using load cells. Axial displacement of the rotating shaft is measured using a linear displacement transducer which provides a measure of sample wear or sample consumption or both during burning. Rotational speed is monitored using an rpm sensor.

The data are digitally processed by a microprocessor and stored on a floppy disk. Data from each instrumentation channel is stored every 100 ms, which represents an averaged value obtained from eight data samples.

Test Procedure

The following procedure was used to perform most of the testing presented in this paper.

1. The test samples were loaded into the chamber and the test pressure was established.
2. The shaft rotational speed was adjusted to obtain the desired sample surface velocity.
3. The test samples were contacted. The contact pressure was steadily increased (1.7×1.0^5 N/m²/s) until either ignition or sample failure occurred or until the maximum contact pressure was obtained.
4. The test samples were examined and observations were recorded.

Alternate procedures included varying the chamber pressure and the sample surface velocity during the test while the other parameters were held constant.

Results and Discussion

The primary objective of the testing described in this paper was to develop a ranking criterion that would measure the resistance of metals to ignition when exposed to frictional heating. Once a ranking criterion was developed, the effects of varying conditions such as surface velocity, contact load, oxygen pressure and so forth, could be determined.

Ignitions were defined as an event that produced a rapid generation of heat and sample consumption. This was detected by a rapid temperature increase (sample thermocouples or optical pyrometer or both), a rapid radiant heat output (thermopile), rapid sample consumption (linear displacement transducer), and a final decrease in the equilibrium oxygen pressure. Final evidence for an ignition was obtained from post-test examination of the test specimens. Figure 7 shows typical responses from the instrumentation for an ignition. Figure 8 shows the appearances of Monel 400 for tests that resulted in a no ignition and an ignition.

Ranking Criterion

The ranking criterion was based on the energy flux required to produce ignition of a particular material. The rate of frictional energy can be expressed by

$$Q_{\text{fric}} = \omega L = P\bar{v}\mu A \tag{3}$$

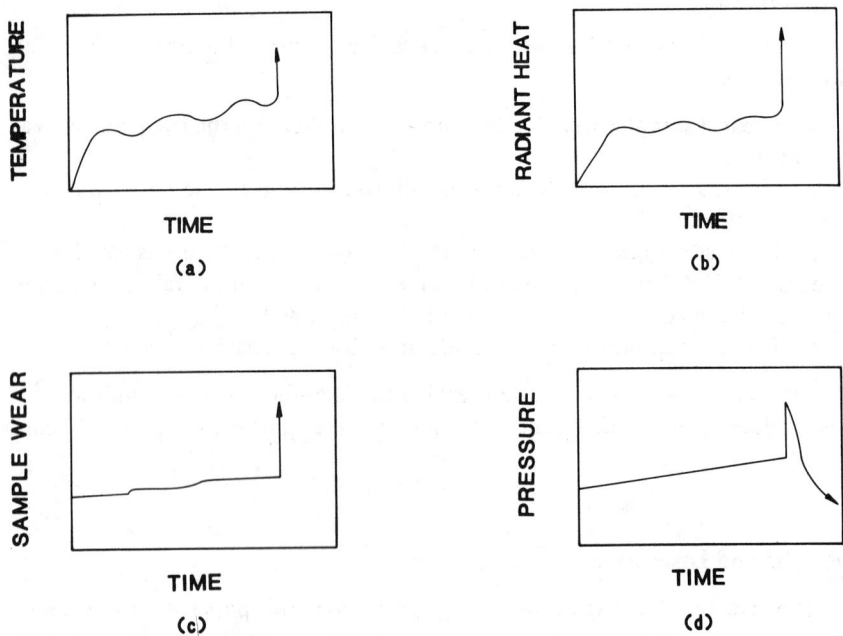

FIG. 7—*Typical response from instrumentation for an ignition event:* (a) *temperature,* (b) *radiant heat output,* (c) *sample wear or consumption,* (d) *oxygen pressure.*

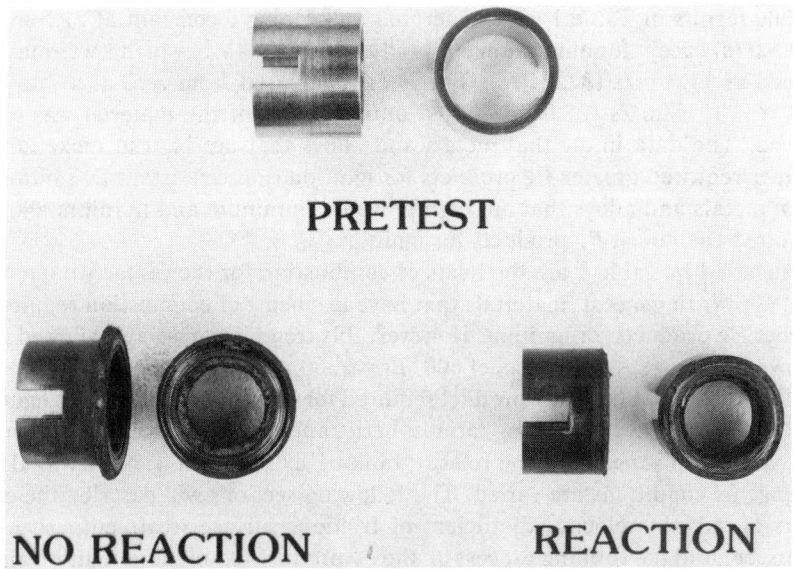

FIG. 8—*Photograph of pre-test and post-test friction rubbing samples: Monel 400.*

Rearranging Eq 3 gives a power term represented by the product of the contact pressure P and the average linear surface velocity $\bar{v}$

$$P\bar{v} = \omega L/A\mu \qquad (4)$$

The units of $P\bar{v}$ product reduce to W/m^2 or $Btu/s/ft^2$ but are more conveniently expressed as N/m^2 m/s or $lbf/in.^2$ in./s. Table 1 gives the ranking of typical metals and alloys with respect to $P\bar{v}$ product required for ignition.

TABLE 1—$P\bar{v}$ *Products required for ignition of typical metals and alloys at 6.9 MPa (1000 psig).*

Material	$P\bar{v}$ Product		Heats of Combustion, kJ/g · mole
	$N/m^2 \cdot m/s \times 10^{-8}$	$(lbf/in.^2 \cdot in./s \times 10^{-6})$	
Nickel 200	2.28 to 3.38	(1.30 to 1.93)	241
Inconel 600	1.99 to 2.89	(1.14 to 1.65)	422
Monel 400	1.42 to 1.55	(0.81 to 0.89)	219
Monel K-500	1.37 to 1.63	(0.78 to 0.93)	260
Hastelloy X	0.93 to 1.26	(0.53 to 0.72)	572
Brass 360	0.69 to 1.17	(0.39 to 0.67)	226
Invar 36	0.60 to 0.94	(0.34 to 0.54)	608
316SS	0.54 to 0.72	(0.30 to 0.41)	802
Aluminum 6061-T6	0.063	(0.036)	1676
Ti-6Al-4V	0.0039	(0.002)	1369

The results in Table 1 were generated by keeping $\bar{v}$ constant at 21.5 m/s (70.5 ft/s) except for aluminum 6061-T6 and Ti-6Al-4V in which $\bar{v}$ was maintained at 11.4 m/s (37.3 ft/s). The P was increased from zero at a rate of 1.7×10^5 N/m²/s (25.0 lbf/in.²/s) until ignition of the material was observed. The data imply that metals and alloys that are high in nickel and copper required greater $P\bar{v}$ products for ignition (more resistant to ignition) than metals and alloys that are high in iron. Aluminum and titanium alloys required the lowest $P\bar{v}$ products for ignition.

Included in Table 1 are the heats of combustion for the respective metals and alloys. In general, materials that have low heats of combustion required higher $P\bar{v}$ products for ignition. However, this trend is not always followed as shown by the results for Inconel 600, Brass 360, and Ti-6Al-4V.

The absolute value of $P\bar{v}$ product required for ignition of a particular material will change as the rubbing, atmospheric, and material surface conditions are varied. In some cases the relative ranking as shown in Table 1 will also change as conditions are varied. The following sections will describe the effects of surface velocity, coefficient of friction, surface treatment, oxygen pressure, and the rubbing process on the Pv product required for ignition and the ignition process in general.

Effect of Surface Speed and Contact Pressure

Figure 9 illustrates the average $P\bar{v}$ products (N/m² · m/s) required for ignition of several alloys as a function of $\bar{v}$ (m/s). The $P\bar{v}$ products were obtained

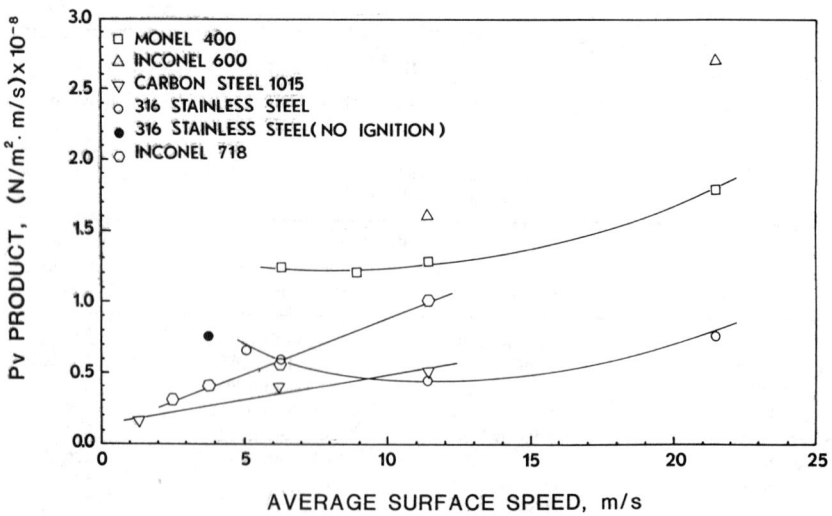

FIG. 9—*Effect of surface speed on P$\bar{v}$ product required for ignition of several alloys at 6.9 MPa (1000 psig).*

by increasing P for a particular $\bar{v}$, which was maintained constant. Each data point presents an average of at least three individual tests and some cases an average of five tests. The results indicate that the $P\bar{v}$ product required for ignition is dependent on $\bar{v}$ and the particular alloy. For example, Inconel 718 and carbon steel 1015 show what appears to be linear increases in $P\bar{v}$ products as $\bar{v}$ increases. On the other hand, Type 316 stainless steel and Monel 400 exhibit nonlinear or nonregular relationships.

The deviation from the linear relationship at the lower $\bar{v}$'s for Monel 400 and Type 316 stainless steel is believed to be caused by the higher P's required to produce ignition at these $\bar{v}$'s. At these high P's, sample deformation was observed. Table 2 gives the average P's required for ignition of several of the materials showed in Fig. 9. In the case of Monel 400 and Type 316 stainless steel, sample deformation occurred at the high P's, which correspond to the lower $\bar{v}$'s in Fig. 9. Sample deformation indicates work was performed on the sample which consumes energy from the rubbing process (see Eq 1; Q_w). Thus a higher $P\bar{v}$ product is required to compensate for this loss of energy. Inconel 718 and carbon steel 1015 did not exhibit sample deformation throughout the P's and $\bar{v}$'s tested, and these materials exhibited linear relationships for $P\bar{v}$ products as a function $\bar{v}$.

In some cases, the relative ranking based on $P\bar{v}$ product for several of the materials changed as $\bar{v}$ was increased. This was most evident with Inconel 718 and Type 316 stainless steel, and carbon steel and Type 316 stainless steel. In

TABLE 2—*Effects of contact pressure (P) on* $P\bar{v}$ *products required for ignition.*

Material	P, N/m² × 10⁻⁷	$P\bar{v}$ Product, N/m² · m/s × 10⁻⁸
Inconel 718	0.89	1.01
	0.97	0.61
	1.13	0.43
	1.24	0.31
Monel 400	0.83	1.79
	1.13	1.29
	1.45	1.21[a]
	1.97	1.24[a]
316 SS	0.36	0.78
	0.47	0.54
	1.08	0.68[a]
	1.31	0.67[a]
	1.97	0.75[b]
Carbon Steel 1015	0.40	0.45
	0.59	0.37
	1.14	0.16

[a]Material exhibited mechanical deformation during rubbing process before ignition.
[b]Material exhibited gross mechanical deformation and did not ignite.

both cases, the change in the relative ranking occurs when sample deformation was observed with the Type 316 stainless steel.

The increase in $P\bar{v}$ product as $\bar{v}$ was increased for the materials shown in Fig. 9 is believed to be caused by an increase in convective cooling and a decrease in the coefficient of friction μ. This is shown in Table 3 for Monel 400 in which P was maintained constant (1.97×10^6 N/m²), and $\bar{v}$ was increased from 6.3 to 21.5 m/s during a particular test run. As $\bar{v}$ was increased, the temperature of the stationary sample (measure 1.3 mm from the rubbing surfaces) decreased. This implies that increased turbulence due to the increase in sample $\bar{v}$ provided a better path for convective heat transfer and thus lowered the sample temperature. Table 3 also shows that the μ decreased as $\bar{v}$ was increased and implies that the frictional energy produced by the rubbing process would also decrease (Eq 3). The end result is that a higher $P\bar{v}$ product is necessary to reach the energy rate required for ignition (Eq 1). However, more testing is required before a definitive conclusion can be made.

Conditions Affecting Coefficient of Friction

The effect of the coefficient of friction μ on $\dot{Q}_{fric}$ is shown by Eq 3; as μ decreases, $\dot{Q}_{fric}$ also decreases. The μ for a particular material has been shown to be dependent upon temperature [9,16], presence of oxygen [9,17], surface velocity [16,17], and loading pressure [17]. The effect of temperature on μ is shown in Fig. 10. The value of μ decreases as the temperature of the sample increases. The effect of oxygen atmospheres as opposed to nitrogen atmospheres on μ are significant. The values of μ were typically two to four times greater in nitrogen than in oxygen.

Table 4 gives the effects of $\bar{v}$ on μ in which P was maintained constant at 8.35×10^6 N/m² and temperature of the stationary sample (measured 1.3 mm from the rubbing surface) was also constant at 472 ± 15 K ($390 \pm 27°$F). The results indicate that μ decreases as $\bar{v}$ increases and supports the results shown in Fig. 9 in which $P\bar{v}$ products required for ignition increased as $\bar{v}$ increased.

TABLE 3—*Results from variable surface speed tests on Monel K-500; contact pressure maintained constant at 1.97×10^6 N/m².*

Surface Speed, m/s	Temperature, K[a]			Coefficient of Friction		
	Low	Average	High	Low	Average	High
6.3	689	714	755	0.15	0.19	0.24
16.4	653	658	700	0.11	0.15	0.17
21.5	608	616	644	0.05	0.14	0.25

[a]Temperature measure on stationary sample 1.3 mm from rubbing surfaces.

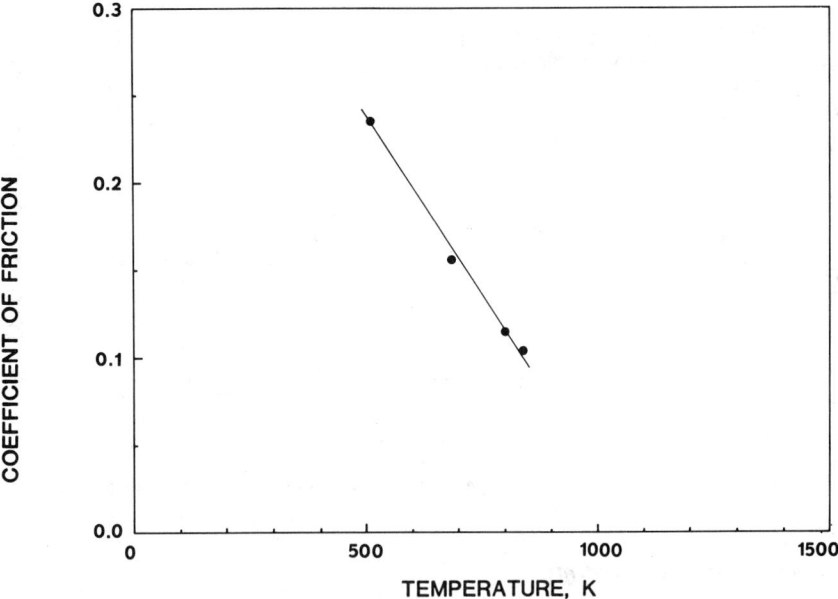

FIG. 10—*Effect of temperature on the coefficient of friction for Monel K-500.*

TABLE 4—*Effect of surface speed on the coefficient of friction for Monel K-500 at a constant P of 8.35 $\times$ 10⁶ N/m².*

Surface Speed, m/s	Average Coefficient of Friction
7.8	0.25
12.6	0.20
16.4	0.15
21.5	0.12

NOTE: Temperature of sample (1.3 mm from rubbing surface) was constant at 472 $\pm$ 15 K.

Tests were conducted to determine the effects of P on the μ. The results are shown in Fig. 11 for Monel K-500 in which P was increased from 2.5 $\times$ 10⁶ to 8.8 $\times$ 10⁶ N/m² at a constant $\bar{v}$ of 21.5 m/s. As P was increased, μ remained fairly constant or slightly decreased at the higher P tested. The literature [17] indicates that μ should increase as P is increased. During these tests, the temperature of the sample increased for each increase in P, and as shown in Fig. 10, μ will decrease for increases in temperature. Thus if an increase in μ was actually occurring in these tests, the increase in temperature overpowered the overall effect of P on μ as measured in these tests.

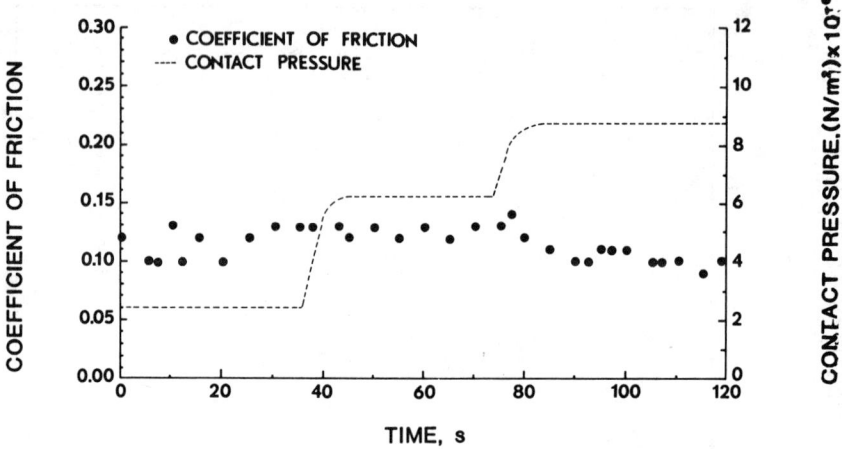

FIG. 11—*Effect of contract pressure on the coefficient of friction for Monel K-500.*

Effects of Surface Treatment

Very little data are available at this time to make any definite conclusions as to the effect of surface treatment, alteration, or doping on the characteristics ignition of metals and alloys. However, the subject will be briefly addressed here because of the importance surface treatment can play in the development of metals and alloys that are more resistant to ignition.

Ignition of metals and alloys by frictional heating is dependent on the surface characteristics in at least two ways. First, frictional energy is developed as the surface and changes in μ can change the rate of frictional energy. Second, surface oxides can affect the oxidation kinetics and thus change the temperature T_c or T_t required for ignition.

Table 5 shows the results for 17-4 PH stainless steel samples that had been either annealed or heat treated (Rockwell Hardness C44). Nearly identical $P\bar{v}$ products were required for ignition at a $\bar{v}$ of 11.4 m/s for the heat treated and annealed samples. At a $\bar{v}$ of 21.5 m/s, the heat-treated samples required higher $P\bar{v}$ products than the annealed samples. The standard deviation of the

TABLE 5—*Aging effects on ignition of 17-4PH stainless steel.*

Material	$P\bar{v}$ Product, $(N/m^2 \cdot m/s) \times 10^{-8}$	
	11.4 m/s	21.5 m/s
17-4 PH (Ann)	0.75 (SD = 0.15)	0.92 (SD = 0.24)
17-4 PH (HT)	0.76 (SD = 0.10)	1.23 (SD = 0.07)

NOTE: Ann: annealed, HT: heat treated to Rockwell Hardness C44, and SD: standard deviation.

$P\bar{v}$ products for the heat-treated samples was much smaller than the standard deviation for the annealed samples. The exact cause for the change in $P\bar{v}$ products are presently being investigated, but the results do suggest that the ignition characteristics of metals and alloys are dependent on their prior conditioning.

Effects of Oxygen Pressure

The effect of increased oxygen pressure on the ignition of carbon steel 1015 is shown in Table 6. The results indicate that as pressure is increased from 0.69 to 20.7 MPa (100 to 3000 psig) the bulk ignition temperature (measured 1.3 mm from the rubbing surface) decreased and the $P\bar{v}$ product required for ignition increased. In theory, increased oxygen pressure can increase the oxidation kinetics and the convective heat loss. The pressure effect on oxidation kinetics is dependent upon the presence of an oxide coating and the type of coating: absorption controlled $(Pox)^{1/2}$ and diffusion controlled $(Pox)^{1/n}$ where n can vary between 5 and 8 [12]. The effect of pressure on laminar free convection and turbulent free convection are approximately $(Pox)^{1/2}$ and $(Pox)^{2/3}$, respectively [18]. As illustrated in Eq 1, the kinetic term will add heat whereas the convective term will remove heat from the rubbing samples. Therefore one would expect that a pressure exists where the absolute values of both terms are equal.

Figure 12 gives the $P\bar{v}$ products required for ignition as a function of oxygen pressure for three alloys. In the case of the carbon steel 1015 and Type 316 stainless steel, there exists a pressure where increasing or decreasing pressure produces increases in $P\bar{v}$ products required for ignition. At the pressure where the minimum $P\bar{v}$ product occurs, it is believed that the heat rate produced by the oxidation process is equal to the heat loss rate. The ignition process at pressures lower than this minimum are dominated by oxidation kinetics whereas at pressure above this minimum, the ignition process is dominated by heat loss from the material. Note that the slopes of the curves for the three alloys are different at the high pressure regions. Since these slopes are the resultant of heat addition and heat loss, they can provide an indication of

TABLE 6—*Bulk ignition temperature and* $P\bar{v}$ *product required for ignition as a function of oxygen pressure (1015 carbon steel).*

Oxygen Pressure, MPa (psig)	Temperature at Ignition Point, K (°F)[a]	$P\bar{v}$ Product, N/m² · m/s × 10⁻⁸
0.69 (100)	1105 (1530)	0.26
3.45 (500)	852 (1075)	0.30
6.89 (1000)	767 (920)	0.33
20.7 (3000)	450 (350)	0.40

[a]Temperature measured at 1.3 mm from the rubbing surface.

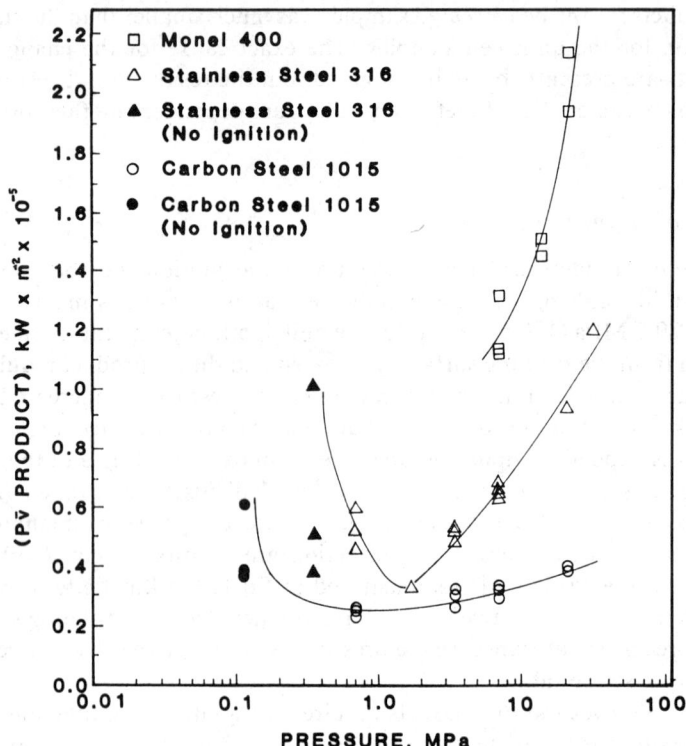

FIG. 12—*Effect of oxygen pressure on the P$\bar{v}$ products required for ignition for three alloys.*

the types of oxidation kinetics involved in the ignition process. However, a model that describes all heat loss paths (conduction, convection, and radiation) must be used to make this determination.

During the pressure evaluation, tests were specifically conducted to insure that convective heat loss was the cause for the increased $P\bar{v}$ products observed at the higher pressure regions. It was postulated that increasing oxygen pressure may cause μ to decrease and thus required higher $P\bar{v}$ products for ignition. In these tests, Monel K-500 samples were rubbed at a constant $\bar{v}$ and P in oxygen initially at 0.69 MPa (100 psig). The samples were rubbed at this pressure until temperature equilibrium of the samples was established. The oxygen pressure was then increased (without changing $\bar{v}$ or P) to 6.9 MPa (1000 psig) and then to 20.7 MPa (3000 psig). The results are given in Fig. 13, which shows the change in sample temperature (measured 1.3 mm from the rubbing surface) and as a function of time. As pressure was increased, sample temperature decreased and μ increased. Similar temperature decreases were predicted using a simple one-dimensional convective heat loss model for

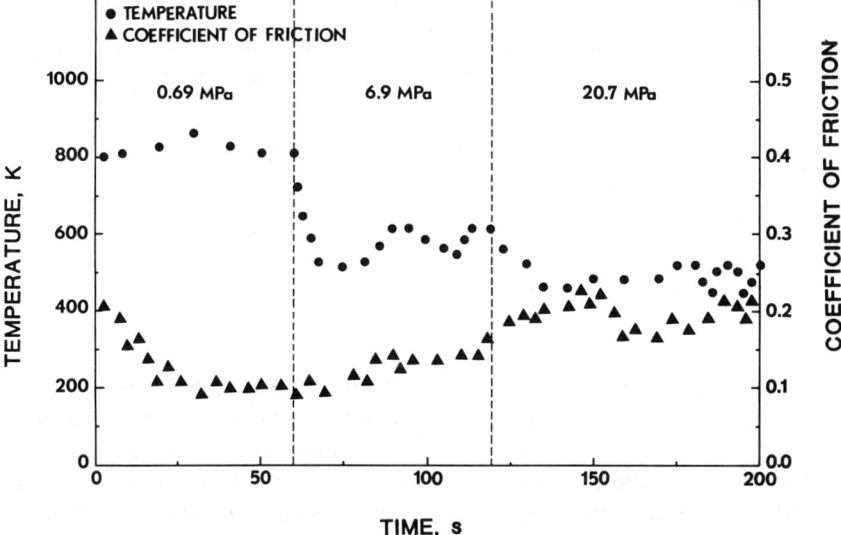

FIG. 13—*Effect of oxygen pressure on sample temperature and coefficient of friction of Monel K-500.*

laminar free convection. The increase in μ as pressure was increased is attributed to the decrease in the sample temperature.

These results indicate that convective heat loss is the major cause for the increase in $P\bar{v}$ product required for ignition as oxygen pressure is increased.

Effects of the Rubbing Process on Ignition

As pointed out in the Background Section, the ignition temperature of metals and alloys is dependent in many cases on the characteristics of the surface oxide coatings that form. The process of rubbing two materials together can provide mechanical stress on the oxide coatings. If the coatings fail, the temperature required for ignition can be different than the ignition temperature obtained, for example, in a heated bomb.

This was evident for the ignition of aluminum, see Table 7. The differences in the ignition temperature obtained in a heated bomb and obtained in a friction rubbing test are significant. Obviously, the oxide coating for aluminum in the friction rubbing test cracked and yielded a lower ignition temperature.

The results for carbon steel and Monel K-500 (see Table 7) indicate that the rubbing process did not significantly affect the ignition temperatures of these two alloys as compared to the ignition temperatures obtained in heated bombs. Metals and alloys that exhibit a decrease in ignition temperature as pressure increases most likely form nonprotective oxide coatings [19]. The

TABLE 7—*Comparison of ignition temperatures obtained in friction rubbing tests and in heated bombs.*

Material	Ignition Temperature[a] (Heated Bomb)		Ignition Temperature (Friction Rubbing)	
	Temperature, K	Pressure, MPa	Temperature, K	Pressure, MPa
Aluminum	2123	6.9	648 to 698	6.9
Carbon steel	1370 to 1450	0.69	1100 to 1310	0.69
Monel K-500	1475	6.9	1370	6.9

[a]Data obtained from Schmidt and Forney [*15*].

ignition temperature of the Monels are reported to be independent of pressure whereas the carbon steels exhibited a decrease in ignition temperature as pressure is increased [*15*]. Thus, the ignition process of the Monels should be characterized by diffusion controlled oxidation and the carbon steels by absorption controlled oxidation. These conclusions agree with those reported by Jenny and Wyssmann [*9*] discussed previously in the Background Section. Therefore, it is expected that the ignition temperature of carbon steel obtained in the friction rubbing test slower similar to that obtained in the heated bomb tests. This indeed is reflected by the data given in Table 7. However, in the case of Monel K-500, the difference in the ignition temperatures obtained by the two difference methods are also similar. Conclusions that can be drawn from these data are (1) Monel does not follow diffusion controlled oxidation, which seems unlikely based on other data, or (2) the friction rubbing process did not significantly affect the oxide coating.

Conclusions

The resistance of materials to ignition by frictional heating in oxygen atmospheres was investigated. A ranking criterion that compares the product of contact pressure P and average surface velocity $\bar{v}$ required for ignition of materials was investigated. Materials that are more resistant to ignition required larger $P\bar{v}$ products than those less resistant to ignition. The absolute values of $P\bar{v}$ products for ignition have been shown to be affected by variations in surface velocity, contact pressure, coefficient of friction, oxygen pressure, and previous conditioning of the material. In some cases, the relative ranking of materials based on $P\bar{v}$ product changed by varying the rubbing conditions. This was shown for Inconel 718 and Type 316 stainless steel by varying surface velocity. Test data have indicated that a more definitive ranking of metals and alloys could be obtained at the high surface velocities because lower contact pressures are required for ignition. Ignition temperatures obtained from friction rubbing tests were compared with ignition temperatures obtained using heated bombs. Of the materials that were compared, aluminum exhibited a significantly lower ignition temperature in the friction rubbing tests than ignition temperatures reported using heated bombs. This reduced ignition temperature of aluminum was attributed to failure of its protective oxide coating because of the rubbing process.

Acknowledgments

The authors wish to acknowledge Randy Shaw, Carl Wright, Ike Melendrez, Charles Murray, and Joe Diaz who conducted the majority of the tests from which the data presented in this paper were obtained. In addition, the authors would like to extend their appreciation to Craig Bishop and Dr. Michael Pedley for their many hours of consultation. The funds for this work has been provided by the Office of the Chief Engineer, NASA Headquarters, Washington, DC, primarily through the efforts of Joyce McDevitt.

References

[*1*] Bates, C. E., Wren, J. E., Monroe, R. W., and Pears, C. D., "'Ignition and Combustion of Ferrous Metals in High Pressure, High Velocity Oxygen," *Journal of Materials for Energy Systems*, Vol. 1, No. 1, June 1979, pp. 61–76.

[2] Naegeli, J. P., "What Triggers Oxygen Turbo Compressor Fires," *Proceedings, Oxygen Compressors and Pumps Symposium, Compressed Gas Association*, Atlanta, GA, 1971, pp. 35–38.

[3] Schoenman, L., "Selection of Burn-Resistant Material for Oxygen-Driven Turbopumps," paper presented at the 20th AIAA/ASME/SAE Joint Propulsion Conference, Cincinnati, OH, 1984.

[4] Clark, A. F. and Hurst, T. G., "A Review of the Compatibility of Structural Materials with Oxygen," *AIAA Journal*, Vol. 12, No. 4, April 1974, pp. 441–454.

[5] Bauer, H., Wegener, W., and Windgassen, K. F., "Fire Tests on Centrifugal Pumps for Liquid Oxygen," *Cryogenics*, June 1970, pp. 241–248.

[6] Bauer, H., Klein, G. K., Wegener, W., and Windgassen, K. F., "Fire Tests on Centrifugal Pumps for Liquid Oxygen—Part 2," *Cryogenics*, June 1970, pp. 241–248.

[7] Lapin, A., "Liquid and Gaseous Oxygen Safety Review," Vol. I–IV. NASA-CR-120922, APCI TM184, NASA Lewis Research Center, Cleveland, OH, June 1972.

[8] Schutt, H. U., Knapp, R. H., and Schmeal, W. R., "Ignition of Some Common Engineering Alloys: The Critical Energy Input Concept and Effects of Oxygen Pressure," *International Corrosion Forum*, National Association of Corrosion Engineers, Houston, TX, March 1976, pp. 14/1–14/23.

[9] Jenny, R. and Wyssmann, H. R., "Friction-Induced Ignition in Oxygen," *Flammability and Sensitivity of Materials in Oxygen-Enriched Atmospheres, STP 812*, American Society for Testing and Materials, Philadelphia, 1983, pp. 150–166.

[10] Stoltzfus, J. and Benz, F. J., "Development of Methods and Procedures for Determining the Ignitibility of Metals in Oxygen," NASA Technical Report TR-281-001 (INT-1), NASA White Sands Test Facility, Las Cruces, NM, Nov. 1984, p. 20.

[*11*] Glassman, I., Mellor, A. M., Sullivan, H. F., and Laurendeau, N. M., "A Review of Metal Ignition and Flame Models," Conference Proceedings No. 52, NATO AGARD Annual Meeting, Feb. 1970, pp. 19(1–30).

[12] Kofstad, P., *High Temperature Oxidation of Metals*, Wiley, New York, 1966.

[13] Hauffe, K., *Oxidation of Metals*, Plenum Press, New York, 1965.

[14] Kubaschewski, O. and Hopkins, B. E., *Oxidation of Metals and Alloys*, 2nd ed., Bullerworth and Co., London, 1967.

[15] Schmidt, H. W. and Forney, D. F., "ASRDI Oxygen Technology Survey, Volume IX: Oxygen Systems Engineering Review," NASA SP-3090, NASA, Washington, DC, 1975.

[16] *Marks Standard Handbook for Mechanical Engineers*, 8th ed., T. Baumeister, Editor in Chief, McGraw-Hill Book Co., New York, 1978.
[17] Adamson, A. W., *Physical Chemistry of Surfaces*. 3rd ed., John Wiley & Son, New York, 1976.
[18] Rohsenow, W. M. and Hartnett, J. P., Eds., *Handbook of Heat Transfer*. McGraw-Hill Book Co., New York, 1973.
[19] Laurendeau, N. M., "The Ignition Characteristics of Metals in Oxygen Atmospheres." Technical Report No. 851, Department of Aerospace and Mechanical Sciences, Princeton University, Princeton, NJ, Oct. 1968, p. 93.

Walter W. Yuen[1]

A Model of Metal Ignition Including the Effect of Oxide Generation

REFERENCE: Yuen, W. W., **"A Model of Metal Ignition Including the Effect of Oxide Generation,"** *Flammability and Sensitivity of Materials in Oxygen-Enriched Atmospheres: Second Volume, ASTM STP 910*, M. A. Benning, Ed., American Society for Testing and Materials, Philadelphia, 1986, pp. 59–77.

ABSTRACT: A theoretical model is presented to simulate the general behavior of the metal ignition process. The model includes many physical mechanisms (such as the various heat transfer rates, reaction rate, and the effect of oxide formation on the reacting surface) that are known to have influence on metal ignition. Based on numerical solutions, three separate types of metal ignition are identified. They are (1) external heating controlled ignition, (2) convection controlled ignition, and (3) convection and transition temperature controlled ignition. For certain types of metal ignition, such as the external heating controlled ignition, the ignition temperature appears to be an ineffective parameter in illustrating the relative ignitability of the different materials. For such cases, the concept of a minimum ignition heat flux is shown to be useful in correlating the relative ignitability.

Test data for five different alloys generated from the friction-rubbing test apparatus at the White Sands Test Facility are analyzed based on the present model. Qualitatively, almost all observed ignitions are shown to be external heating controlled. Tests with lower rotational speed and higher oxygen pressure are needed to determine the minimum ignition heat flux for the five alloys. Based on the current set of data, some conclusions concerning the minimum ignition heat flux of Type 316 SS and a preliminary ranking of the five alloys in terms of their relative ignitability are generated. These conclusions, however, require further verification by additional test data.

KEY WORDS: metals, heat transfer, ignition temperature, ignition, oxide layer, minimum ignition heat flux, metal ranking

The ignition of metal in an oxygen environment is a problem of considerable practical importance in the aerospace engineering community. Over the years, many experimental and theoretical studies [1–6] have been conducted attempting to understand the various physical mechanisms governing the ignition process. Most of these studies, however, are limited to isolated mecha-

[1]Associate professor, Department of Mechanical Engineering, University of California, Santa Barbara, CA 93106.

nisms and simple systems. Because of the general complexity of the ignition process, the application of most of these results to practical engineering systems has been quite limited. In the selection of materials for aerospace application, for example, the current procedure still utilizes standardized tests (such as the mechanical impact test) [7-9], which are known to generate inconsistent ranking results.

The objective of this work is to develop a general theoretical model simulating the metal ignition process. The model will include many physical mechanisms (such as the various heat transfer rates, reaction rate, and the formation of oxide at the reacting surface) that are known to have effects on metal ignition. In contrast to many previous theoretical models [1,2,6], the emphasis of the present model will not be on the detailed quantitative study of one particular physical mechanism or experiment. Instead, the model will simulate the interaction of the different mechanisms and demonstrate their relative importance in the general ignition process. The model will be kept sufficiently general so that it can be applied to interpret data obtained from different types of ignition test.

Over the past two years, a friction-rubbing test apparatus has been designed and constructed at the White Sands Test Facility (WSTF). Some limited ignition data for 316 stainless steel (SS), Hastelloy Alloy X, Invar-36, Monel-400, and Monel K-500 have recently been obtained. The present model will be used to analyze these data and to generate some conclusions concerning the relative ignitability of the different alloys. Based on the data analysis, limitations of the existing friction-rubbing test apparatus as a standardized test for metal/alloy screening will be noted. Some suggestions concerning future tests to be conducted with the apparatus will be made.

Theoretical Model and Predictions

General Assumptions and Equations

When a metal surface is exposed to an oxygen environment, it is well known [10,11] that a reaction will occur with a rate of reaction that depends strongly on the metal temperature and the thickness of the oxide layer formed at the reacting surface. Based on the so-called Evans mechanisms [10,11], the oxidation law can be written in the following form

$$(dM/dt) = Ae^{-\delta/\delta_o}e^{-E/RT} \tag{1}$$

where M is the total mass of metal reacted per unit reacting area, t is time, δ is the thickness of the oxide film accumulated on the reacting surface, A and E are the rate constant and activation energy for the reaction that can be functions of temperature and pressure, R is the universal gas constant, and T is the surface temperature. The term $e^{-\delta/\delta_o}$ characterizes the effect of retarded oxidation caused by the presence of an oxide layer on the reacting surface. The value δ_o determines the general order of magnitude of the retardation effect. It varies with different metals/alloys and must be determined by a detailed consideration of the oxide formation process and properties of the oxide layer. The factor $e^{-E/RT}$ characterizes the temperature dependence of the oxidation. It is important to note that there exists a large class of oxidation laws different from Eq 1. Equation 1 is chosen because it is mathematically simple, and it illustrates the essential physics of oxidation retardation caused by increased oxide film thickness. The use of a different oxidation law is not expected to lead to any significantly different qualitative conclusions from those generated by the present work.

Based on Eq 1, the energy equation for a piece of metal subjected to a heat flux $H(t)$ in an oxygen environment is given by

$$\frac{\rho V C}{S}\frac{dT}{dt} = QAe^{-\delta/\delta_o}e^{-E/RT} - h(T - T_\infty) - \epsilon\sigma(T^4 - T_\infty^4)$$

$$- \frac{1}{R_i}(T - T_i) + H(t) \qquad (2)$$

where ρ is the metal density, V/S is the effective volume to reacting surface area ratio, C is the specific heat, σ is the Stefan-Boltzmann constant, Q is the heat generated per gram of metal reacted, h is the external heat transfer coefficient, ϵ is the metal's emissivity, T_∞ is the ambient temperature, R_i is an internal resistance to heat transfer, T_i is the corresponding internal reference temperature, and $H(t)$ is the external heat input to the metal. For the metal oxide accumulating over the reacting surface S a mass conservation equation can be written as

$$\rho_{ox}\frac{d\delta}{dt} = \gamma Ae^{-\delta/\delta_o}e^{-E/RT} - D(t) \qquad (3)$$

where ρ_{ox} is the density of the oxide, γ is the ratio of the mass of oxide generated to the mass of metal reacted, and $D(t)$ is an oxide removal rate that might vary for the different situations to which the model is applied. Equations 2 and 3 are the basic governing equations describing the heat and mass transfer at the metal surface before ignition.

It is important to note that while Eqs 2 and 3 are written for a metal at uniform temperature T, they can be applied to more complex situations provided the different parameters $(T, V, S, h, \epsilon, (E/R), Q, A, R_i,$ and $T_i)$ are interpreted appropriately. If T is interpreted as an average temperature and S an effective area that accounts for the difference between surface temperature and average temperature, for example, Eq 2 is an accurate description of energy conservation for a metal with nonuniform temperature distribution. Differences between reacting area and heat transfer area can also be accounted for with an appropriate definition of h. In short, Eqs 2 and 3 can be applied to describe general metal ignition behavior.

Introducing the following dimensionless variables

$$\vartheta = \frac{RT}{E}, \qquad \tau = \frac{QASRt}{\rho CEV}, \qquad \eta = \frac{\delta}{\delta_o}$$

$$H'(\tau) = \frac{H(t)}{QA}, \qquad D'(\tau) = \frac{D(t)}{\gamma A}$$

$$h_c = \frac{hE}{QRA}, \qquad h_\tau = \frac{\epsilon \sigma E^4}{QAR^4}$$

$$\lambda = \frac{\rho CEV}{QSR}\left(\frac{\gamma}{\rho_{ox}\delta_o}\right), \qquad h_i = \frac{E}{R_i QAR}$$

and

$$\vartheta_\infty = \frac{RT_\infty}{E}, \qquad \vartheta_i = \frac{RT_i}{E}$$

the following nondimensional conservation equations result

$$\frac{d\vartheta}{d\tau} = e^{-\eta}e^{-1/\vartheta} - h_c(\vartheta - \vartheta_\infty) - h_\tau(\vartheta^4 - \vartheta_\infty^4) - h_i(\vartheta - \vartheta_i) + H'(\tau) \qquad (4)$$

$$\frac{d\eta}{d\tau} = \lambda[e^{-\eta}e^{-1/\vartheta} - D'(\tau)] \qquad (5)$$

Physically, ϑ can be interpreted as a ratio of the temperature to a characteristic reaction temperature (E/R). $[(\rho CEV)/(QASR)]$ is the characteristic time required to heat the metal to the characteristic reaction temperature E/R with the heat of reaction Q. τ is the ratio of the actual time to that characteristic time. QA is the characteristic heat flux generated by the reaction. $H'(\tau)$, h_c, h_τ, and h_i are dimensionless ratios representing the relative importance between external heat input, convective heat transfer, radiative heat transfer, and internal heat transfer to the heat transfer caused by the reaction. γA can be interpreted as a characteristic oxide generation rate. The ratio $(\rho_{ox}\delta_o/\gamma A)$ is the characteristic time required to generate an oxide layer of thickness δ_o per unit reacting surface area. The dimensionless parameter λ can thus be interpreted as a ratio of the characteristic time for heat generation to that for mass generation.

Characteristics of Metal Ignition without Oxide Layer

Assuming that there exists an external mechanism that prevents the accumulation of oxide layer on the reacting surface, Eq 4 is solved numerically
(with $\eta = 0$) to illustrate the characteristics of metal ignition without the effect of oxide layer. For a constant h_c and ϑ_∞ and setting $h_r = h_i = 0$ (that is,
no radiant and internal heat transfer), the metal temperatures for different
H' are shown in Fig. 1.

For cases in which ignition occurs, it can be readily observed that depending on the external heating rate, the metal ignition process can be classified
into two distinct types. They are

1. External Heating Controlled Ignition—In this case, the external heat-
ing effect $[H'(\tau)]$ is large compared to that of convective cooling $[h_c(\vartheta - \vartheta_\infty)]$
and reactive heating ($e^{1/\vartheta}$). The temperature profile is essentially identical to
that generated by only external heating until ignition occurs. The temperature profile labelled "HC" (heating controlled, $H'/H_o = 15$) in Fig. 1 is an
example of this class of ignition.

2. Convection Controlled Ignition—In this case, the external heating effect is of the same order as that of convective cooling. The metal temperature
rise is much slower than that generated by only external heating until ignition
occurs. The temperature profile labelled "CC" (convection controlled,
$H'/H_o = 1.5$) in Fig. 1 is an example of this class of ignition.

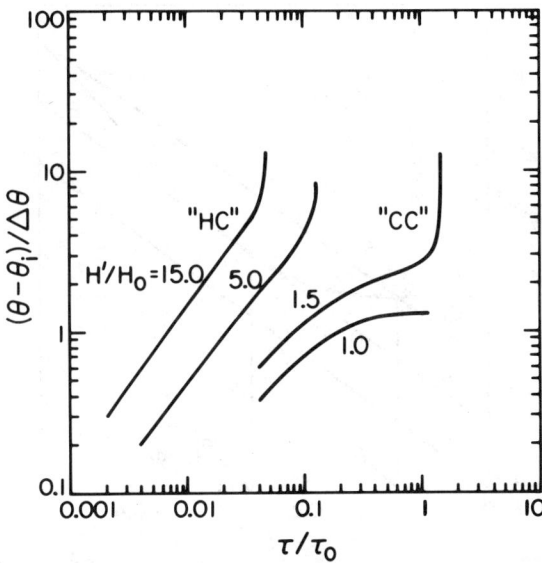

FIG. 1—*Characteristic temperature rise for a metal under a constant external heat flux and a
constant convective heat transfer coefficient* ($h_o = 7.5e - 24$, $h_c = 2.e - 20$, $\tau_o = 4.e + 20$,
$\theta_i = 0.018$, $\Delta\theta = 3.e - 4$).

Results in Fig. 1 also show that as the external heating rate decreases, the convective cooling effect can become dominant and, consequently, suppress ignition. Mathematically, the condition under which ignition is suppressed can be readily established. As in most conventional studies of metal ignition [2,8], the heat addition to the metal can be separated into two components Q_r and Q_l as follows

$$Q_r = e^{-1/\vartheta} \tag{6}$$

and

$$Q_l = h_c(\vartheta - \vartheta_\infty) - H' \tag{7}$$

Physically, Q_r is the heat addition caused by chemical reaction while Q_l is the heat addition (or loss) caused by convection and other external mechanisms. The mathematical behavior of Q_r and Q_l as functions of ϑ for different heat addition rates H' is illustrated in Fig. 2. Curve A represents the case of small heat addition and nonignition. The metal is maintained at an equilibrium temperature ϑ_A at which $Q_r = Q_l$. Curve C represents the case of runaway combustion since Q_r is greater than Q_l over all the temperature range. The highest possible equilibrium temperature $\vartheta_{e,\max}$ and the corresponding minimum ignition heat flux $H'_{\min}$, before this runaway combustion occurs, is illus-

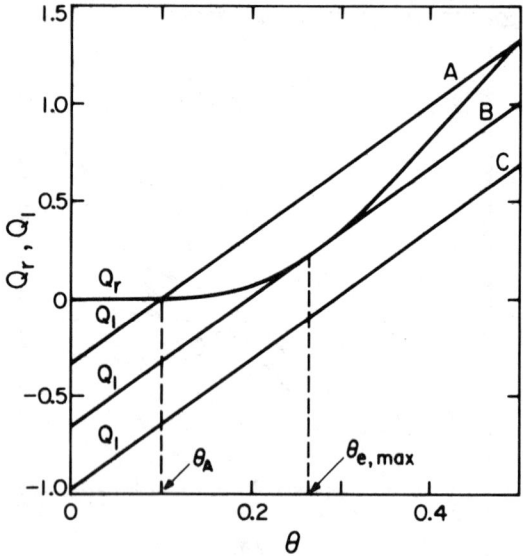

FIG. 2—*Comparison between* Q_l *and* Q_r *for a nonignition case* (A), *ignition case* (B), *and a run-away combustion case* (C).

trated by Curve B. Numerical values for $\vartheta_{s,\max}$ and $H'_{\min}$ can be readily generated from solutions to the following equations

$$\frac{dQ_r}{d\vartheta} = \frac{dQ_l}{d\vartheta} \tag{8}$$

$$Q_r = Q_l \tag{9}$$

They are presented in Figs. 3 and 4.

It is interesting to note that historically, many researchers have attempted to characterize metal ignitability with the determination (both experimentally and theoretically) of an "ignition temperature." Reynolds, for example, interpreted $\vartheta_{s,\max}$ as calculated in Fig. 3 as an ignition temperature [2]. An excellent review of the different definitions of ignition temperature was given by Mellor [6]. Both Mellor's review and the current results, as presented in Fig. 1, show that the precise determination of an ignition temperature is quite difficult, particularly for situations in which the ignition is external heating controlled. Different theoretical definitions, for example, would yield different values for the ignition temperature. Even if a common theoretical definition can be agreed upon, different heat input rates would lead to different values for the ignition temperature. For example, if an ignition temperature

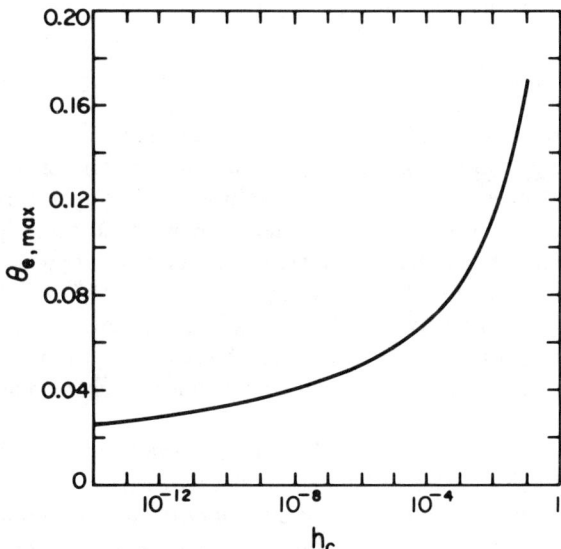

FIG. 3—*Maximum equilibrium temperature as a function of the convective heat transfer coefficient.*

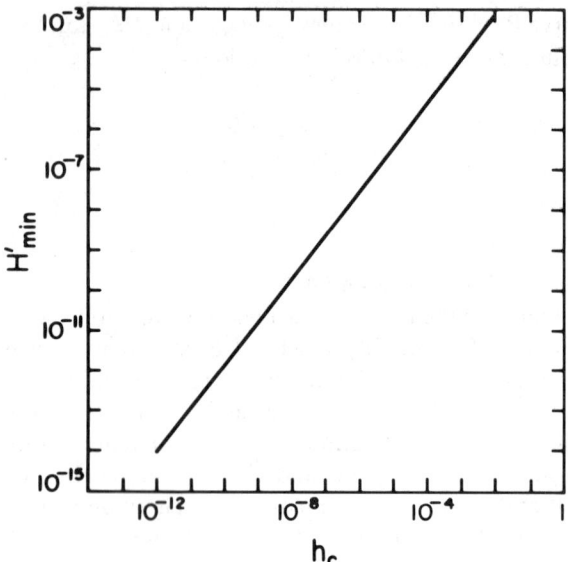

FIG. 4—*The minimum required heat flux for ignition as a function of the convective heat transfer coefficient.*

ϑ_{ig}, is defined as the temperature at which the rate of temperature increase is twice the external heating rate, that is

$$(d\vartheta/d\tau)_{\vartheta=\vartheta_{ig}} \geq 2H' \qquad (10)$$

values of ϑ_{ig} for different H' are generated from the numerical solutions and presented in Fig. 5. Since complex mathematical relations, such as Eq 10 are often quite sensitive to even small uncertainty in the temperature data, ignition temperature is a difficult parameter to measure. Indeed, ignition temperature is often taken to be the last measured temperature before the breakdown of the thermocouple caused by combustion. It can thus also be depended on the sensitivity of instruments. In short, ignition temperature is not an effective parameter to characterize the relative ignitability of different metals/alloys when external heating is the primary mechanism leading to ignition.

The minimum ignition heat flux H'_{min}, on the other hand, is only a function of h_c. Results in Fig. 1 show that a small change in H' around H'_{min} ($H'_{min} = 1.1H_o$ for cases presented in Fig. 1) is sufficient to change a nonignition case to an ignition case. H'_{min} should thus be a readily measurable parameter. Physically, it is also reasonable to identify a metal/alloy with a low H'_{min} as more ignitable. The minimum ignition heat flux is thus an effective parameter to characterize the relative ignitability of different metals/alloys.

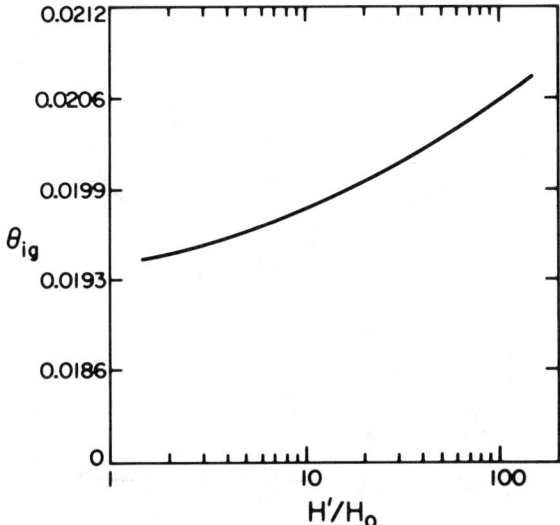

FIG. 5—*Ignition temperature as a function of the constant external heat flux* ($H_o = 7.5e -$ *24, and* h_c *is assumed to* 2.e $-$ *20 in the calculation*).

For a linearly increasing heat input rate, $H'(\tau) = \beta\tau$, the metal temperature histories before ignition for different rate constants β are presented in Fig. 6. Similar to cases with constant heating rate, the ignition process can be identified as either external heating controlled (cases with a large rate of increase in heat input, large β) or convection controlled (cases with a small rate of increase in heat input, small β). Since the heating rate increases monotonically, the metal ignites for all values of β. Using Eq 10 as the definition of ignition, the heat input rate at ignition, $H'_{ig} = H'(\tau_{ig})$, is tabulated for different β and presented in Fig. 7. It is interesting to note that H'_{ig} decreases with decreasing β and approaches H'_{min} as β approaches zero. The concept of minimum ignition heat flux is thus applicable even in ignition process generated by a nonconstant heat input rate.

Characteristics of Metal Ignition with Oxide Layer

To illustrate the effect of oxide layer, numerical solutions to Eqs 4 and 5 are generated with different values of λ. For cases with no oxide removal ($D' = 0$) and constant external heat input, the effects of oxide formation on a convection controlled ($H'/H_o = 1.5$) and an external heating controlled ignition ($H'/H_o = 5.0$) are illustrated in Figs. 8 and 9. As expected, the formation of oxide layer retards the reaction rate and suppresses the ignition.

Physically, an oxide layer cannot remain indefinitely protective under all physical conditions. Even when there is no mechanical means to remove the

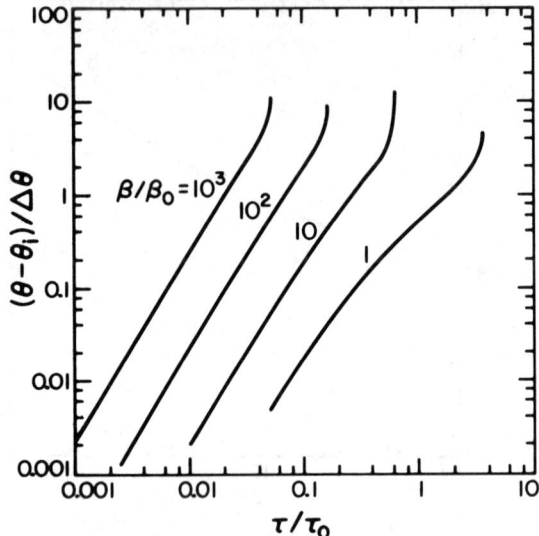

FIG. 6—*Characteristic temperature rise for a metal under a linearly increasing heat flux (*H'* = β τ) and a constant convective heat transfer coefficient. (β₀ = 8.5e − 45, all other parameters are as defined in Fig. 1).*

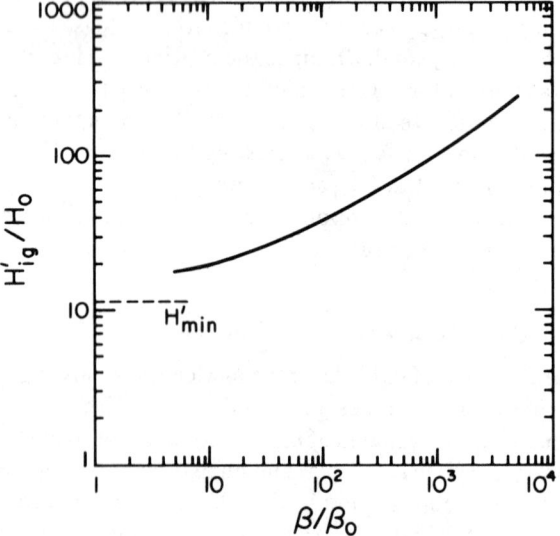

FIG. 7—*The calculated ignition heat flux for different values of β (β₀ is as defined in Fig. 6).*

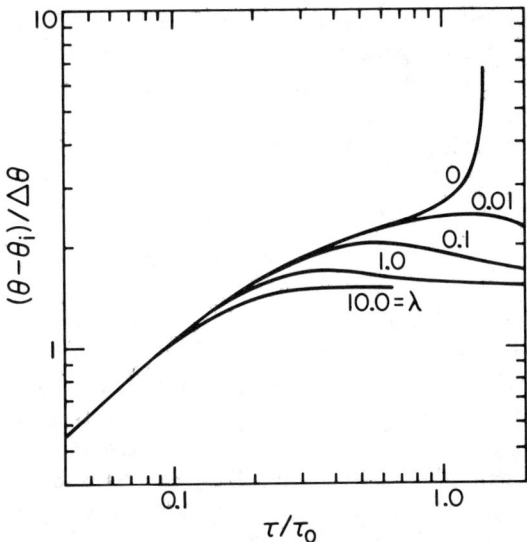

FIG. 8—*Characteristic temperature rise of a metal heated externally by a constant heat flux* $(H'/H_o = 1.5)$ *with different oxide generation rates (H_o and other parameters are identical to cases considered in Fig. 1).*

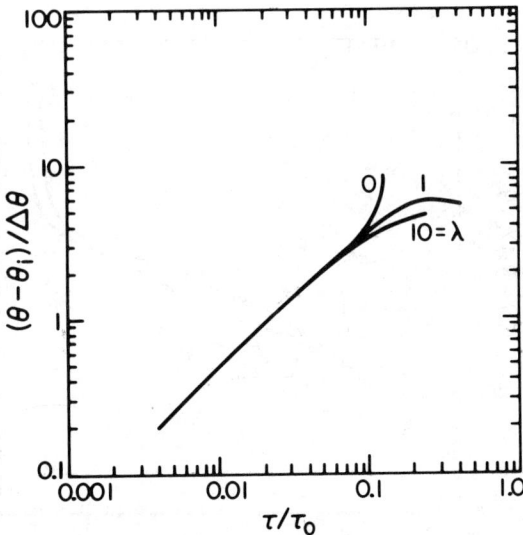

FIG. 9—*Characteristic temperature rise of a metal heated externally by a constant heat flux* $(H'/H_o = 5.0)$ *with different oxide generation rates (H_o and other parameters are identical to cases considered in Fig. 1).*

oxide layer, an oxide layer can lose its protective properties at a temperature called the transition temperature. The general concept of the transition temperature and its physical interpretation were first introduced by Mellor [6]. Mathematically, the concept of the transition temperature ϑ_t can be approximated by the following definition for the oxide removal rate D'

$$D'(\vartheta) = 0 \qquad \vartheta < \vartheta_t$$
$$\qquad\qquad \infty \qquad \vartheta > \vartheta_t \qquad\qquad (11)$$

For the same external heat inputs as considered in Figs. 8 and 9, the combined effects of oxide formation and a finite transition temperature are illustrated in Figs. 10 and 11. For the external heating controlled case, it is interesting to note that if the transition temperature is sufficiently low such that ignition occurs, the resulting temperature history is almost indistinguishable from that of an externally heating controlled ignition without the formation of oxide layer. For the convection controlled ignition, however, the temperature history is qualitatively different from that with no oxide generation. Indeed, the temperature history labelled as CTTC (convection and transition temperature controlled ignition) in Fig. 10 can be classified as a third general type of metal ignition. The initial temperature rise is slower than that generated by external heating alone because of convection heat loss and the reaction retar-

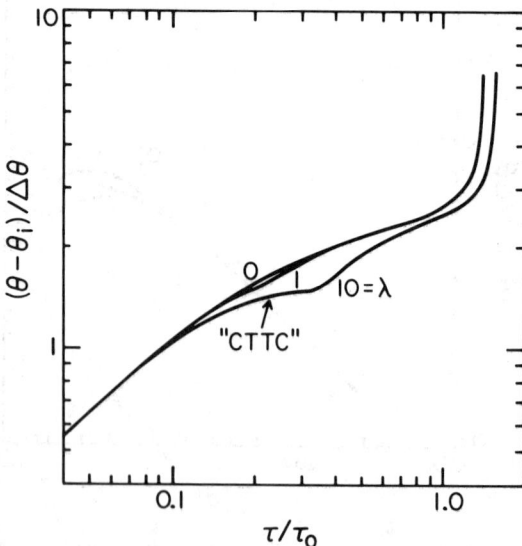

FIG. 10—*Characteristic temperature rise of a metal with a finite transition temperature* $((\theta_t - \theta_i)/\Delta\theta = 1.6)$ *and different oxide generation rates under constant external heating* ($H'/H_o = 1.5$). *(All other parameters are identical to those in Fig. 1.)*

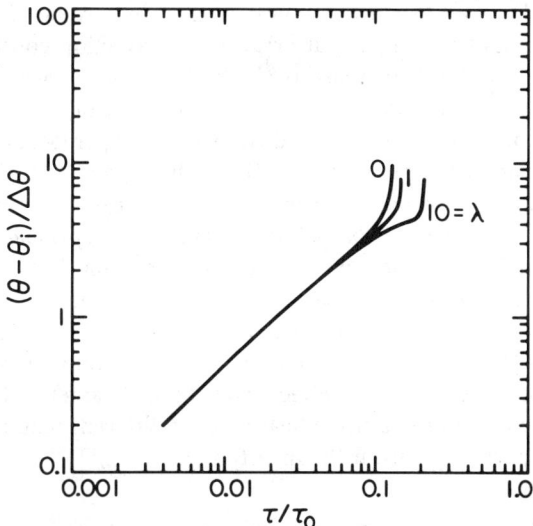

FIG. 11—*Characteristic temperature rise of a metal with a finite transition temperature [(θ_t − θ_i)/$\Delta\theta$ = 4.5] and different oxide generation rates under constant external heating (H'/H$_o$ = 5). (All other parameters are identical to those cases considered in Fig. 1.)*

dation effect of the oxide layer. At the transition temperature, the rate of temperature increase rises sharply leading to ignition.

Because of the presence of an oxide layer, the external heat input required for ignition is higher than that for cases with no oxide layer. While the required heat input for ignition is now a complicated function of h_c, β, λ, and ϑ_t, it is higher than the minimum ignition heat flux calculated for the same h_c. The concept of minimum ignition heat flux as a metal-ranking criterion thus remains valid.

Summary of Theoretical Conclusions

Two important conclusions are generated from the present theoretical model. They are

1. Most metal ignition processes can be classified into three general types based on the characteristics of the metal temperature rise before ignition. The three general types of metal ignition are

(a) external heating controlled ignition,
(b) convection controlled ignition, and
(c) convection and transition temperature controlled ignition.

In an external heating controlled ignition, the rate of temperature increase is governed entirely by the external heating rate until ignition occurs. Formation of oxide layer and the transition temperature have no effect on the general characteristics of temperature rise. In a convection controlled ignition, the rate of temperature increase is slower than that generated by external heating. The convective heat loss slows down the rate of temperature increase before ignition. In a convection and transition temperature controlled ignition, the transition temperature is sufficiently large so that both convective heat loss and oxide formation decrease the rate of temperature rise. At the transition temperature, the temperature increases sharply leading to ignition.

2. For ignition processes in which external heating is a primary heating mechanism, the concept of ignition temperature is not an effective metal ranking parameter. It is difficult to determine experimentally and can also be a function of external conditions. The concept of minimum ignition heat flux, however, appears to be a more effective ranking parameter. In many testing situations, the minimum ignition heat flux for different metals/alloys can be determined accurately with little uncertainty.

Analysis of WSTF Friction Rubbing Test Data

Background

The detail of the test procedure and facilities are described elsewhere.[2] Five different alloys were tested at an identical oxygen pressure of 69 bar (1000 psig). Type 316 SS was tested at different oxygen pressures. The heat input is generated by rubbing the face of a hollow rotating cylinder against the face of a hollow stationary cylinder with the applied load increasing linearly at 31.1 N/s (7 lbf/s). Different heat input rates were generated by varying the speed of the rotating cylinder.

In terms of the qualitative results on the ignitability of a particular alloy at a certain oxygen pressure and cylinder rotational speed, the tests were quite repeatable. Three separate tests, for example, were conducted for 316 SS at identical oxygen pressure (69 bar, 1000 psig) and cylinder rotational speed (17 000 rpm). Ignition was observed in all three cases. But in terms of quantitative data, such as surface temperature measurements, the agreement between the different tests with "identical" test conditions is not good. At the time right before ignition, difference in temperature data as high as 200 K were observed. The difficulty in obtaining repeatable temperature measurements can probably be attributed to the many uncontrolled parameters that existed in the experiment.

In view of the quantitative uncertainty of the temperature data and the general lack of information on the reaction constant A and activation energy E for the different alloys, the present work will not analyze the data quantitatively. Instead, qualitative trend will be identified. Together with the theoreti-

---

[2]Benz, F. J. and Stoltzfus, J. M., in this publication, pp. 38–58.

cal model, limited conclusions regarding the ignitability of the different alloys will be established.

316 SS Data

Temperature data obtained for Type 316 SS at an oxygen pressure of 69 bar (1000 psig) and different rotational speeds (5000, 9000, and 17 000 rpm) are presented in Figs. 12a, b, and c. For a constant friction coefficient of 0.1, the temperature rise caused by friction heating alone is calculated and shown as the dotted curves on the three figures. In spite of the large uncertainty of the temperature data, it can be readily observed that for the cases of 9000 and 17 000 rpm, the measured temperature history does not deviate significantly from the predicted temperature rise due only to friction heating, they can thus be considered as external heating controlled ignitions. In the case of 5000 rpm, however, the measured temperature drops below the predicted temperature generated by friction heating before ignition occurs. This behavior is similar to a convection controlled ignition as defined in Fig. 1. But in view of

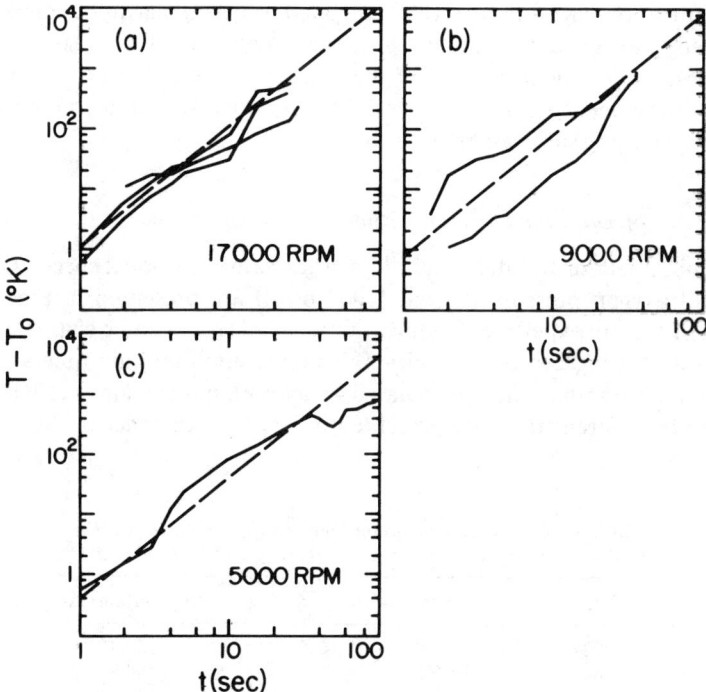

FIG. 12—*Temperature data for 316 SS at an oxygen pressure of 69 bar (1000 psig) and different rotational speeds. (The dotted curve is the temperature increase generated by external heating only.)*

the general large uncertainty of the temperature data, more tests are required before any general conclusions can be made concerning the characteristic of ignition at different rotational speeds.

Based on the measured time of ignition and the assumption of a constant friction coefficient of 0.1, the ignition heat flux for the different tests are tabulated and shown in Table 1. While the ignition heat flux definitely decreases as the rotational speed decreases from 17 000 to 9000 rpm, it increases at 5000 rpm. Theoretically, the ignition heat flux is expected to decrease and approach the minimum ignition heat flux asymptotically as the rotational speed decreases. The discrepancy between theory and experiment can probably be attributed to the general experimental uncertainty. The friction coefficient should also decrease with decreasing rotational speed.

Temperature data obtained at a fixed rotational speed of 5000 rpm and different oxygen pressures are presented in Figs. 13a through d. The dotted curves again represent temperature rise because of external heating with a friction coefficient of 0.1. Based on comparisons between the measured temperature histories and the predicted temperature rise caused by friction heating, cases with lower oxygen pressures (6.9 and 34.5 bar [100 and 500 psig]) appear to be external heating controlled ignition while cases with higher oxygen pressures (69 and 207 bar [1000 and 3000 psig]) appear to be convection controlled. This general trend can probably be attributed to the increase in heat transfer coefficient because of the increase in oxygen pressure. Again, more data are needed before a more definitive conclusion concerning the effect of oxygen pressure can be generated.

Comparison of Data for Different Alloys

Data obtained for five different alloys at the same rotational speed (17 000 rpm) and oxygen pressure (69 bar [1000 psig]) are presented in Figs. 14a through e. Within experimental uncertainty, all of them can be characterized as external heating controlled ignition. The measured ignition time and the estimated ignition heat flux (calculated by assuming a friction coefficient of 0.1) for the different tests are presented in Table 2. In order of decreasing

TABLE 1—*The measured ignition time and the estimated ignition heat flux for 316 SS at an oxygen pressure of 69 bar (1000 psig).*

Test	rpm	t_{ig}, s	H_{ig}, cal/cm$^2 \cdot$ s
22	5 000	97.8	202.4
84	9 000	29.0	108.2
85	9 000	35.6	132.8
140	17 000	24.2	170.3
141	17 000	28.3	199.2
142	17 000	25.3	178.9

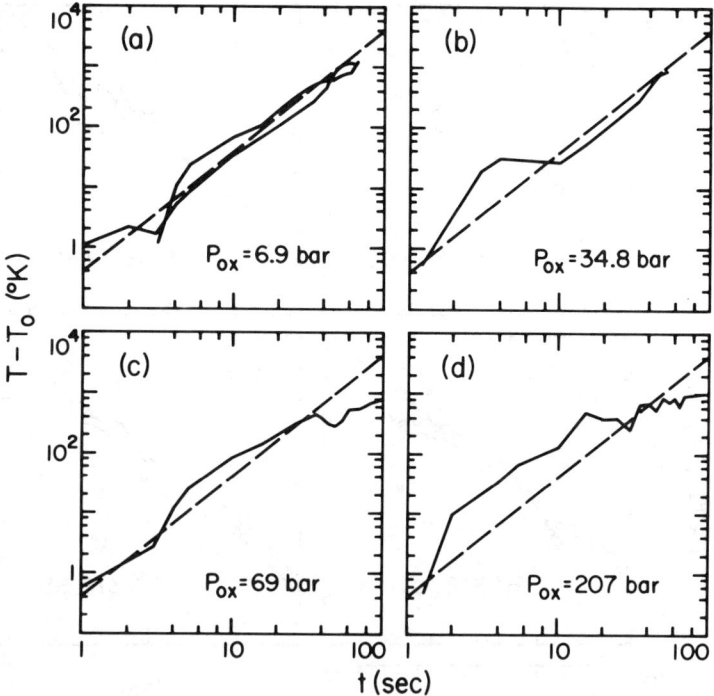

FIG. 13—*Temperature data for 316 SS at 5000 rpm with different* P_{ox}. *(Dotted curve is temperature increase caused by external heating only.)*

ignition heat flux, the five alloys can be ranked as follows: Monel-400, Monel K-500, Hastelloy X, 316 SS, and Invar-38. It is important to note that the above ranking is not based on comparison of the minimum ignition heat flux. It is thus uncertain and subject to further experimental verification.

General Conclusions on WSTF Data

The following conclusions are generated from analysis of the existing WSTF data on 316 SS, Hastelloy X, Invar-36, Monel-400, and Monel K-500:

1. The current set of test data consists almost entirely of external heating controlled ignition. Additional tests should be conducted at lower rotational speed and higher oxygen pressure so that convection controlled ignition can be observed and the minimum ignition heat flux for the different alloys can be determined.

2. It is difficult to estimate the effect of oxide formation based on the current set of data because as noted in the theoretical section, the presence of oxide layer and the transition temperature has only small effect on the tem-

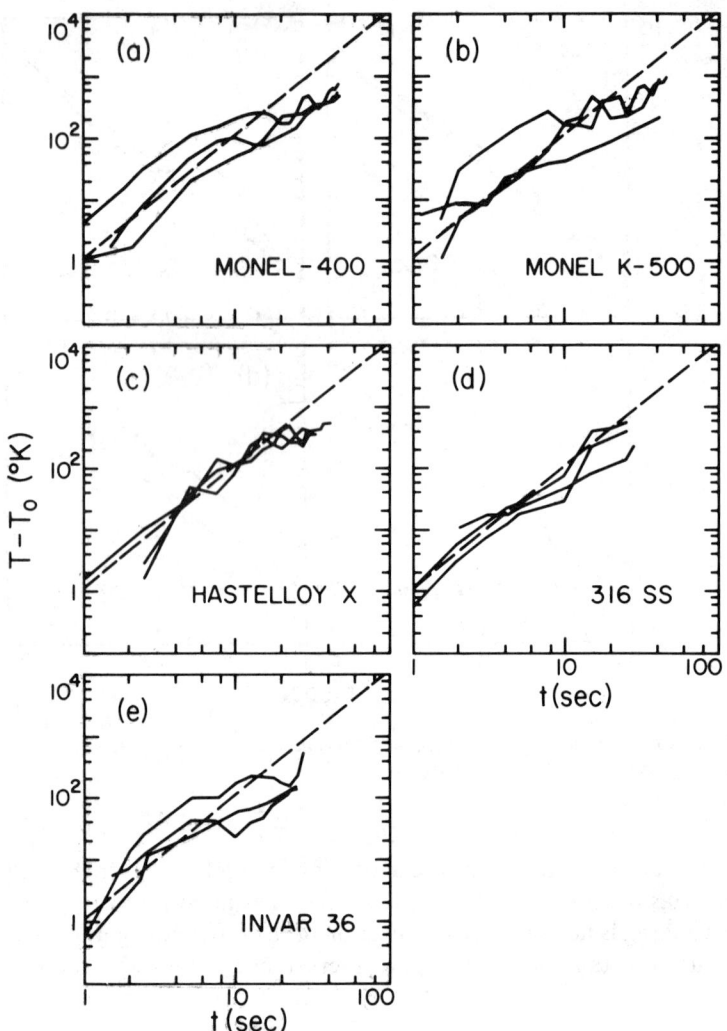

FIG. 14—*Temperature data for five different alloys at 17 000 rpm and 69-bar (1000-psig) oxygen pressure. (Dotted curve is the temperature increase generated by external heating only.)*

perature history of an external heating controlled ignition. Additional data at lower heating rate will help to understand more clearly the effect of oxide generation in a friction-rubbing test.

3. Based on the current set of data, the best estimate for the minimum ignition heat flux of 316 SS at an oxygen pressure of 69 bar (1000 psig) and the environmental conditions of the WSTF friction-rubbing test facility is in the range of 108 to 202 cal/(cm$^2 \cdot$ s).

TABLE 2—*The measured ignition time and the estimated ignition heat flux for different alloys at 17 000 rpm and an oxygen pressure of 69 bar (1000 psig).*

Test	Alloy	t_{ig}, s	H_{ig}, cal/cm^2 · s
162	Monel-400	45.5	320.2
163	Monel-400	46.9	330.1
164	Monel-400	47.7	335.7
179	Monel K-500	45.8	322.3
180	Monel K-500	41.2	290.0
181	Monel K-500	41.3	290.7
151	Hastelloy-X	32.4	228.0
152	Hastelloy-X	29.2	205.5
153	Hastelloy-X	41.5	292.1
140	316 SS	24.2	170.3
141	316 SS	28.3	199.2
142	316 SS	25.3	178.9
149	Invar-36	24.8	174.5
150	Invar-36	25.7	180.9
154	Invar-36	27.7	195.0

4. In order of decreasing ignition heat flux, the five alloys tested at the WSTF friction rubbing facility at an oxygen pressure of 69 bar (1000 psig) and a rotational speed of 17 000 rpm can be ranked as follows: Monel-400, Monel K-500, Hastelloy X, 316 SS, and Invar-36.

5. Both Conclusion 3 and 4 are preliminary and should be further verified with additional test data.

References

[1] Grosse, A. V. and Conway, J. B., *Industrial and Engineering Chemistry*, Vol. 50, No. 4, 1958, p. 668.

[2] Reynolds, W. C., "Investigation of Ignition Temperature of Solid Metals," NASA TN-D-182, Oct. 1959.

[3] Littman, F. E., Church, F. M., and Kinderman, E. M., *Journal of Less-Common Metals*, Vol. 3, 1951, p. 367.

[4] Aldushin, A. P., Bloshenko, V. N., and Seplyarskii, B. S., *Combustion, Explosion and Shock Waves*, Vol. 4, No. 9, July–Aug. 1973, p. 489.

[5] Markstein, G. H., *AIAA Journal*, Vol. 1, No. 3, March 1963, p. 550.

[6] Mellor, A. M., "Heterogeneous Ignition of Metals: Model and Experiment," Ph.D. thesis, Princeton University, Princeton, NJ, 1968.

[7] Lucas, W. R. and Riehl, W. A., *ASTM Bulletin*, 1960, p. 29.

[8] Pippen, D. L. and Stradling, J. S., *Material Research and Standards*, Vol. 11, June 1971, p. 35.

[9] Clark, A. F. and Hust, J. G., *AIAA Journal*, Vol. 12, No. 4, April, 1974, p. 441.

[10] Kubaschewski, O. and Hopkins, B. E., *Oxidation of Metals and Alloys*, 2nd ed. Academic Press Inc., London, Butterworths, 1962.

[11] Hauffe, K., *Oxidation of Metals*, Plenum Press, New York, 1965.

James W. Bransford[1]

Ignition and Combustion Temperatures Determined by Laser Heating

REFERENCE: Bransford, J. W., **"Ignition and Combustion Temperatures Determined by Laser Heating,"** *Flammability and Sensitivity of Materials in Oxygen-Enriched Atmospheres: Second Volume, ASTM STP 910*, M. A. Benning, Ed., American Society for Testing and Materials, Philadelphia, 1986, pp. 78–97.

ABSTRACT: A laser heating technique and facility have been developed to study metal ignition and combustion in high-pressure oxygen. The ignition and combustion temperatures, estimates of oxidation rates, and ignition and combustion morphology can be determined. This facility and the laser heating techniques are described. Examples of the type of data obtained are presented and discussed. The ignition temperature curves for an aluminum alloy—Unified Numbering System (UNS) A96061, a stainless steel—UNS S30200, and two nickel alloys—UNS N07718 and N04400 are given.

KEY WORDS: alloys, aluminum, combustion, ignition, metals, nickel, steels

Sudden and sometimes unexplained combustion of the metallic materials in equipment containing liquid and gaseous oxygen has occurred sporadically for many years. The extent of the combustion damage has ranged from almost undetectable to major system failures. Many of these combustion events involved metallic materials that were thought to be highly resistant to ignition and combustion.

During the development of the Space Shuttle Main Engine (SSME), a number of metal combustion events have occurred. These events ranged from small combustion spots in cracked piping, found only after thorough inspection, to costly oxygen pump failures. Because of the failures during the SSME program, it was decided that a better understanding of the metal ignition and combustion process should be attempted and data generated for specific alloys.

[1]Research chemist, National Bureau of Standards, Center for Chemical Engineering/Chemical Engineering Science Division, Boulder, CO 80303.

To achieve this understanding of the metal ignition and combustion process and to characterize a series of alloys, a program was initiated at the National Bureau of Standards (NBS) by the National Aeronautics and Space Administration (NASA) George C. Marshall Space Flight Center (MSFC). The alloys that are to be characterized are used in the SSME but have applications in other high-pressure oxygen systems.

The metal ignition and combustion characterization process may be approached in two ways. The first is a fundamental approach in which the molecular species and processes are defined and the controlling parameters are identified and quantified. Obviously, to obtain this information at the temperatures and pressures encountered in this program is difficult at best. To obtain information at more workable conditions and to extrapolate to accident conditions is also difficult. Thus, the second approach, to determine general ignition and combustion parameters, has been employed initially. This procedure, it is hoped, will simplify the determination of the fundamental processes that control the ignition and combustion process. This approach will also allow the generation of information for material evaluation purposes, which is a critical part of the program.

The purpose of this paper is to describe the experiments being undertaken in the NBS/MSFC program and to present results for several materials.

Experimental Description

Facility Configuration

The general facility configuration was in large part dictated by the types and number of measurements to be made and by the heat source. Since general combustion parameters were to be obtained, the measurements consisted of the following:

(1) surface and interior specimen temperature for determining the metal oxide-metal interface temperature,

(2) change in specimen mass for oxidation rate behavior,

(3) specimen morphology during the experiment–combustion type, surface behavior, and so forth, and

(4) pressure.

The heating source selected was a continuous wave (cw) carbon dioxide (CO_2) laser. This source allowed a minimum of combustible material to be used in both the specimen and the chamber internal support structure, a major safety consideration, and does not interfer with the chemistry or physics of the specimen surface as, for example, a chemical ignitor would.

There were two major problems in using the laser as a heat source. The first was the nonuniformity of the laser beam. At low pressures, one or more large hot spots can be generated, depending on beam mode. However, this problem was greatly reduced at high pressures because of constant random beam re-fraction from gas convection currents rising from the specimen surface. En-ergy spikes (plasma noise) were also superimposed on the primary beam mode but did not appear to cause hot spot problems because of the random, rapidly changing position of these spikes on the specimen surface. The second problem area was beam power control. Lasers are inherently nonlinear in their control parameters. This required the procedure of setting the power and allowing the specimen to heat at its own rate. However, the latest laser employed in this program does have plasma tube current control and a power monitor now being employed. In the future, a computer-controlled heating ramp may be implemented.

Up to three temperature measurements were made. These were

(1) interior specimen temperature by a W5%Re versus W26%Re (no longer used) and a Pt versus Pt10% Rh thermocouple,
(2) integrated average surface temperature by two-color pyrometry, and
(3) central spot (0.5 mm diameter) surface temperature by two-color py-rometry.

The thermocouple, made from 0.125-mm (0.005-in.) diameter wire, was placed in a 0.625-mm (0.025-in.) hole drilled axially in the specimen to within 0.25 to 0.40 mm (0.010 to 0.015 in.) of the top surface. The temperature mea-sured by the thermocouple was considered to reflect the metal oxide-metal interface temperature more accurately than either surface temperature mea-surement because of the thick, partially insulating nature of the oxide layer at the ignition temperature. The timing of the ignition and combustion temper-ature selection was derived from the integrated average surface temperature waveform. This temperature was a surface average measurement and thus reflected all events that cause temperature changes on the surface; that is, hot spots, phase changes, micro-combustions, and so forth. The spot tempera-ture measurement was abandoned because of effects of laser beam refraction and plasma noise on the data.

The mass of the specimen was monitored by a linear voltage differential transformer (LVDT) type force transducer. This device provided data on the oxidation and combustion rate of the specimen. The mass waveform also pro-vided confirmation of the start of combustion.

Information about specimen morphology was provided by two instruments: a rapid scanning spectrometer and a high-speed camera. The spectrometer can scan a range of 400 to 1100 nm, or a subdivision of this range, in as little as 10 ms; it was used to detect vapor combustion emissions and to determine combustion temperatures. The high-speed 16-mm camera, with framing rates up to 10 000 fps, was used for direct specimen cinematography. Schlieren and Hilbert transform photography can also be performed but was not implemented.

The uncalibrated brightness of the specimen was measured by a silicone photodiode. The waveform from this device has been used in the past to determine the start of combustion but was used only as a backup in case of pyrometer failure in our studies. The sensitivity of the photodiode to small area temperature changes cannot match the sensitivity of the integrated average temperature signal from the two-color pyrometer; thus the pyrometer signal was used to determine timing of ignition and combustion events. The photodiode signal simply was used as an event trigger source.

The experiments were carried out in the high-pressure chamber shown in Fig. 1. Eight ports were provided; four equally spaced at 90° to the chamber axis, and four equally spaced at 45° to the chamber axis. Each port had a clear diameter of 2.75 cm (1.125 in.). The chamber was protected from burning material and the reflected laser beam by one to four layers of a fiber mat heavily impregnated with an alkaline earth silicate cement. The focused laser beam entered the chamber through a sodium chloride window located in the top closure. The experiment was observed through sapphire windows which were protected from hot material by a thin quartz disk. Chamber pressure was measured by a bourdon tube pressure gauge.

All electronic data were recorded by digital oscilloscopes using analog-to-

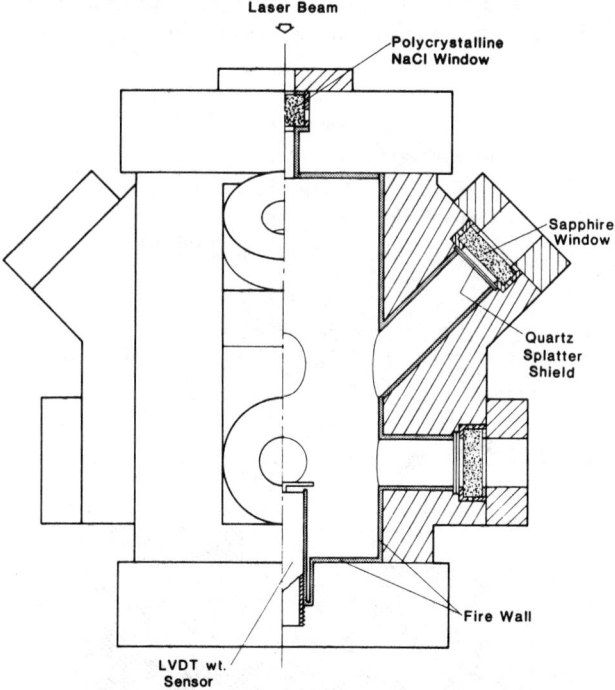

FIG. 1—*Pressure chamber used in metals combustion experiments.*

digital converters with either 12-bit resolution, 50 μV, or 15-bit resolution, 3.125 μV. Spectral data were also photographically or digitally recorded. All digital data was processed by computer and stored on magnetic disk as well as on magnetic tape as backup.

Experimental Procedure

The experimental setup is shown in Fig. 2. The mass sensor, which served as a mounting platform for the experiment, was insulated from the high temperatures generated by the laser beam and combustion by a low-density pad, machined from a porous fire brick. The graphite block, with a machined specimen chamber, served as a containment vessel and massive heat sink. The 5-mm-diameter by 5-mm-high specimen (now changed to 6.35 by 6.35 mm) was insulated from the graphite block by 2 to 4 mm of fine aluminum oxide powder packed to a density of approximately 20% of solid aluminum oxide.

The experimental setup was assembled by first inserting a silicate-aluminum oxide insulated thermocouple through the bottom of the graphite block and weakly cementing it in place. The thermocouple wires were routed through a twin hole alumina insulator, which was placed between grooves in the graphite block and insulating pad. The alumina insulator was then cemented in place. Aluminum oxide powder was packed in place and the speci-

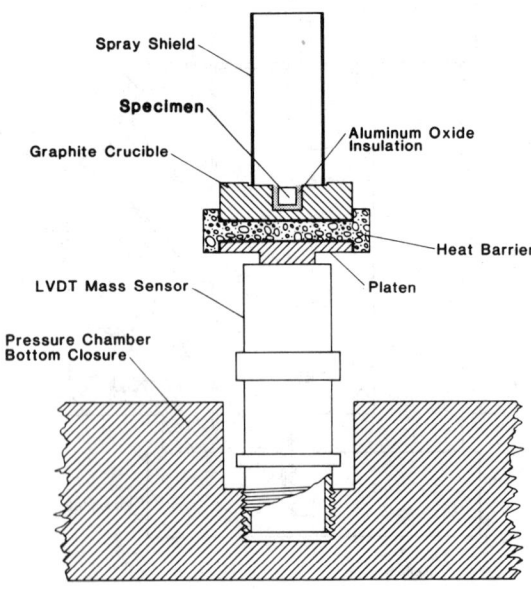

FIG. 2—*Experimental assembly.*

men inserted over the thermocouple. Sufficient force was applied to break the thermocouple-graphite block bond and allow the specimen to rest on the insulating powder. The assembly was placed on the mass sensor platen and the thermocouple connected. The mass sensor was calibrated, chamber closed, evacuated, and pressurized to the desired pressure.

Results and Discussion

The metallic materials that are of interest in the NBS/MSFC program have two types of combustion behavior: vapor combustion and surface combustion. The characteristic waveforms produced by each type of combustion and the ignition temperatures for several materials are discussed in separate sections below. The definitions of the terms that will be used follow:

1. *Ignition Point*—The position on a waveform time axis, usually the integrated average surface temperature waveform, at which a sustained combustion zone or combustion was considered to have developed.

2. *Ignition Temperature*—The temperature determined from a waveform at the ignition point.

3. *Combustion Point*—The position on a waveform time axis, usually the integrated average temperature waveform, at which combustion, unsustained by an external heat source was considered to have developed.

4. *Combustion Temperature*—The temperature determined from a waveform at the combustion point.

Vapor Combustion

The rapidity of the development of combustion was the single most predominant characteristic associated with vapor combustion. Sharp steps were generated in any electronic signal detecting phenomenon associated with ignition and combustion. Figure 3 presents a typical integrated average surface temperature and mass waveform for an experiment. The brightness and spot temperature waveforms are not shown since these waveforms were seldom used. An interior temperature waveform is not shown since the ignition temperatures are greater than the melting point of the Pt versus Pt10% Rh thermocouple, and the W5Re-W26Re thermocouple cannot be used at oxygen pressures greater than 0.34 MPa (50 psia) because of the high surface oxidation rate of the materials.

Aluminum and aluminum alloys were the predominant vapor burning materials studied in the NBS/MSFC program. Of these, the alloy A96061, has been the most extensively tested material to date. Several very interesting aspects of the ignition process have been observed for this alloy. These aspects are shown in the integrated average surface temperature waveform shown in Figs. 4 through 6. Figure 4 shows progressive ignition to combustion se-

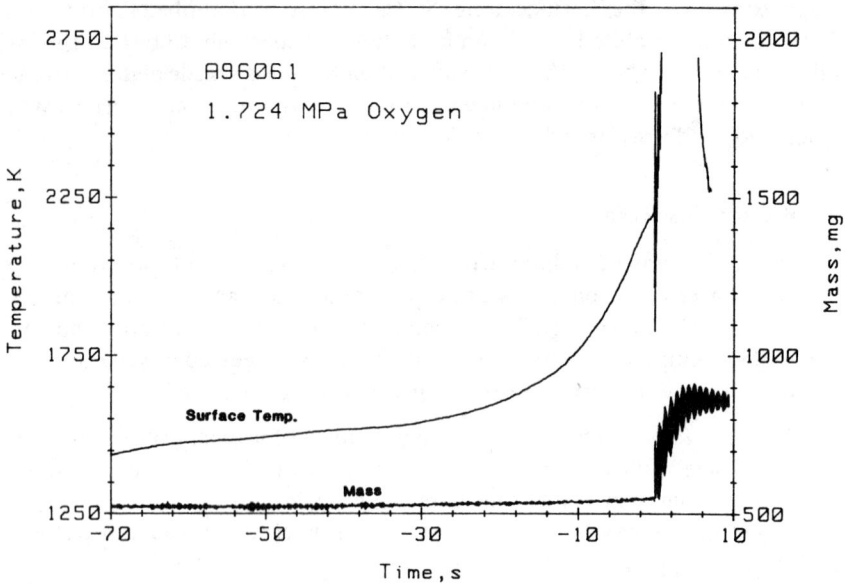

FIG. 3— *Typical surface temperature and mass waveforms for an A96061 combustion experiment.*

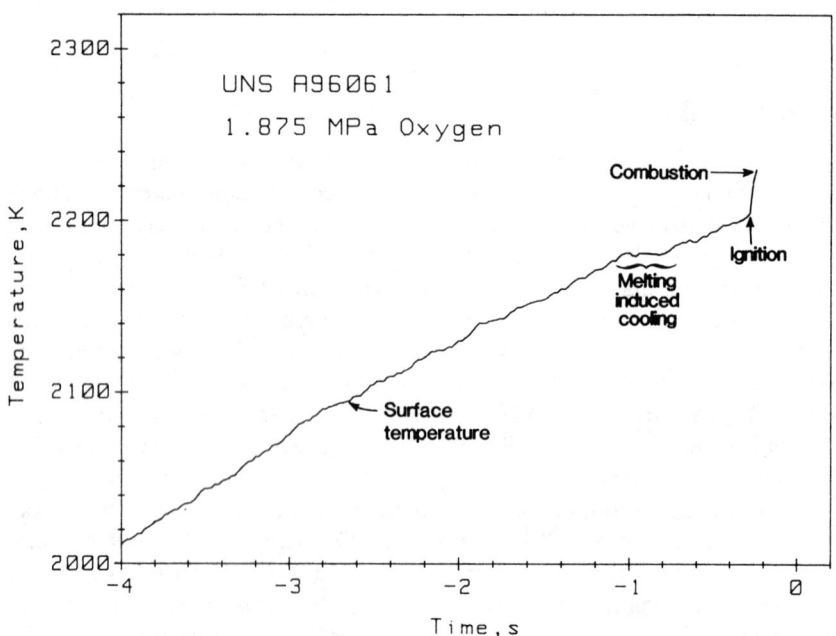

FIG. 4— *Accelerating temperature ignition sequence.*

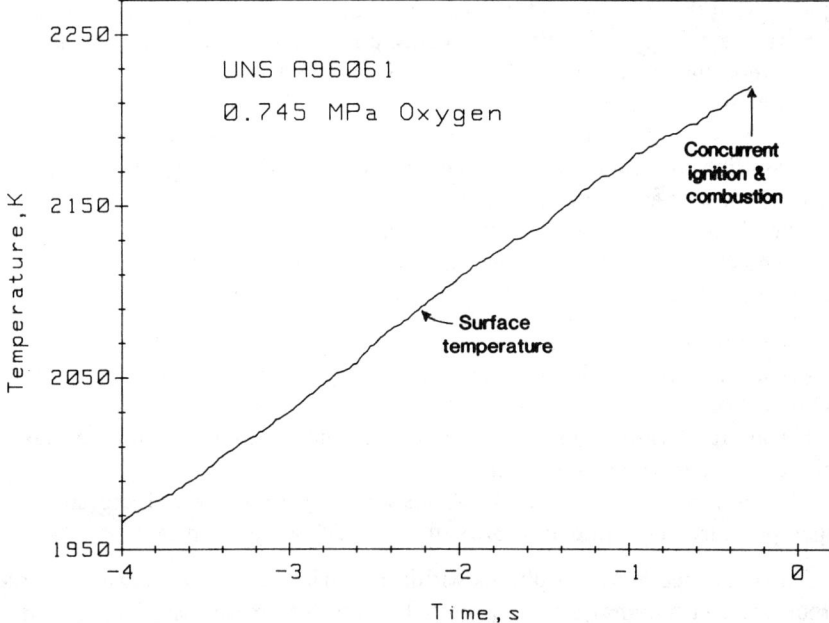

FIG. 5—*Abrupt ignition sequence.*

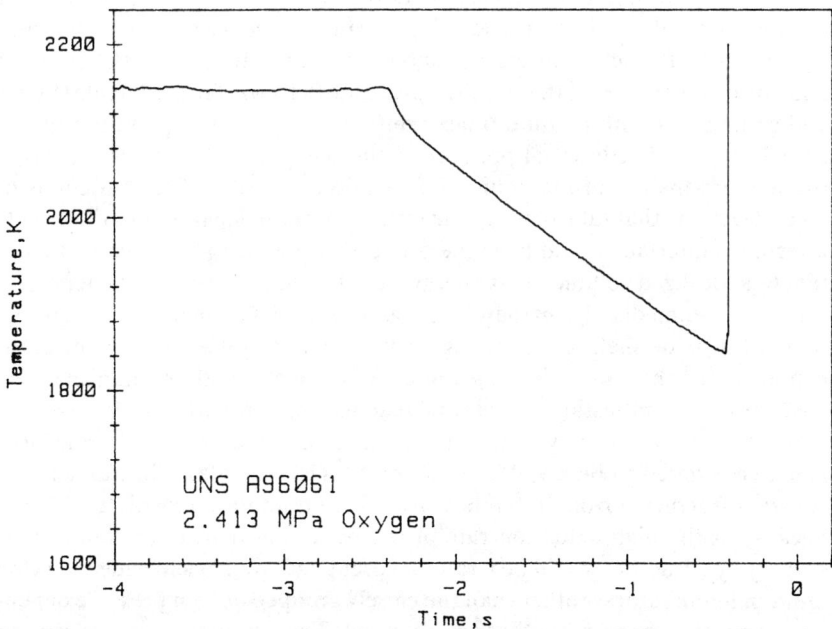

FIG. 6—*Abrupt ignition during specimen cooling.*

quence that develops over a time period of 100 ms. The accelerating temperature rise of this type of ignition sequence almost always ended in a temperature step that exceeded 30 000 K/s. Figure 5 shows a step ignition to combustion sequence, that is, the increase in surface temperature exceeded 30 000 K/s. A small phase change preceded both events. Figure 6 shows a step ignition to combustion sequence that developed as the specimen cooled after the laser beam was turned off.

The exact ignition mechanism(s) that initiated combustion during heating is unknown at this time. However, any mechanism must include a rationale to explain the following observed events:

(1) small spot ignition point,
(2) a step function ignition sequence–temperature risetime greater than 30 000 K/s,
(3) an accelerating ignition sequence terminating in a step function temperature rise to combustion, and
(4) a decrease in ignition and combustion temperatures with increasing oxygen pressure beginning at approximately 0.207 MPa (30 psia).

The existence of liquid phases within the oxide shell, a conclusion based upon the thermal analysis of reproducible waveform features (Figs. 4, 8, and 9), and the linear form of the decrease in ignition temperature, within the limits of present data, suggested a solubility mechanism transporting oxygen to the alloy-alloy oxide interface. The abruptness of the ignition sequence or abrupt termination of the accelerating ignition sequence suggested the sudden contact of reactive species; probably molten aluminum with reducible alloying element oxides, that is, chromic oxide (Cr_2O_3), silicon dioxide (SiO_2), and so forth, or with a liquid phase containing significant quantities of dissolved oxygen. The internal pressure of the specimen, because of the larger thermal expansivity of the liquid alloy, could be a cause of the sudden contact. Reactions that take place in the interior of the oxide layer or at the oxide layer-alloy interface would have the property of retaining the energy of reaction in a localized volume. This energy would go into increasing the temperature of the immediately surrounding material, probably melting a small section of the oxide shell or vaporizing a small volume of molten alloy, or both, rupturing the shell and initiating combustion. One or both mechanisms may have initiated combustion if the initial reactions(s) were sufficiently energetic.

Ignition caused entirely by the development of a fracture in the oxide layer is not considered to be a viable mechanism. The oxide layer has undergone constant fracturing from initial heating. These fractures immediately "heal" because of the high oxidation rate of the liquid alloy. Also ignition caused entirely by a fracture would be a random event and would yield widely varying ignition temperatures rather than the closely grouped oxygen pressure dependent ignition temperatures that are observed. There is one possible exception to the above opinion. A propagating fracture could be the source of the accelerating ignition sequence if a mechanism exists that can initiate the fracture at reproducible temperatures at a given oxygen pressure. As yet, no such mechanism has been formulated.

The mechanism by which ignition was achieved upon specimen cooling appears to be more straightforward. This mechanism is thought to be an external pressure generated implosion or rupture of the oxide surface caused by the larger thermal contraction of the molten alloy with respect to the oxide shell. The resultant sudden exposure of the molten alloy to oxygen, in a geometry that would largely preclude efficient heat loss, should generate sufficient heat to raise the exposed alloy surface to the boiling point or melt the resultant oxide film or both thus allowing combustion to develop. In addition to exposing the molten aluminum to oxygen, the implosion would also place reactive metal oxides into contact with molten aluminum, generating additional energy.

Figure 7 presents the integrated average surface temperature at ignition as a function of oxygen pressure for A96061. It should be pointed out again that the temperature is an integrated average surface temperature and not the exact surface temperature of the area where the ignition developed but is believed to be close to the true temperature. This temperature should be

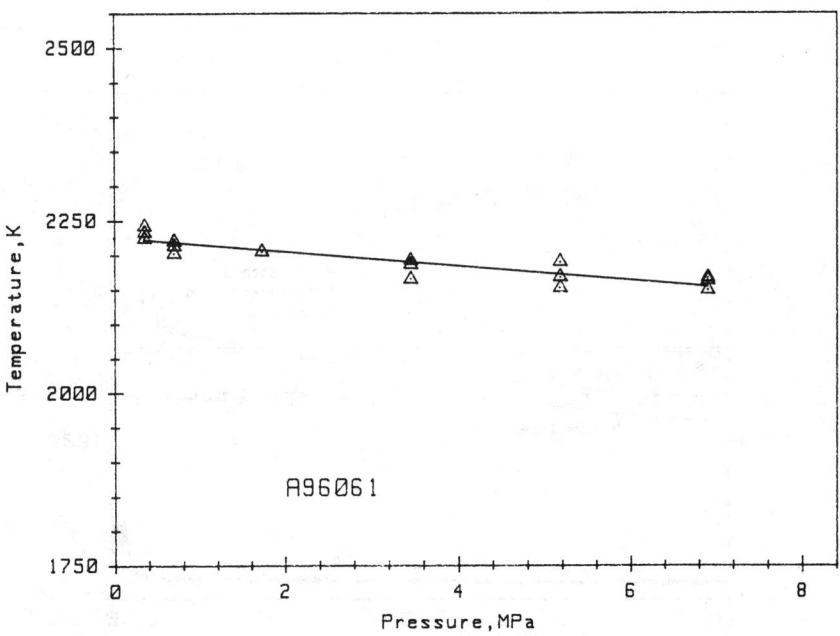

FIG. 7—*Variation of average surface temperature at ignition with oxygen pressure for A96061.*

thought of as a narrow temperature range in which the ignition occurs. This range decreased with increasing oxygen pressure at the rate of 10.3 K/MPa.

Surface Combustion

The bulk of materials that are being characterized in the NBS/MSFC program do not support vapor combustion but burn (rapid surface oxidation) as a liquid. These materials are iron, nickel, and cobalt based alloys. The ignition and combustion mechanism(s) for these materials can be complex; however, certain similarities existed among many of the materials being characterized. The principal similarities that have been observed are

(1) the physical form of the oxide layer that forms during heating and the effect of this layer on the temperature data,
(2) the ignition sequence, and
(3) the combustion sequence.

Within each of the above similarities several variations existed. Figures 8 through 13 present sets of experimental waveforms for three alloys that show the typical similarities and variations in material behavior. The solidus Ts and liquidus $T1$ temperatures for the respective alloy are given as a reference.

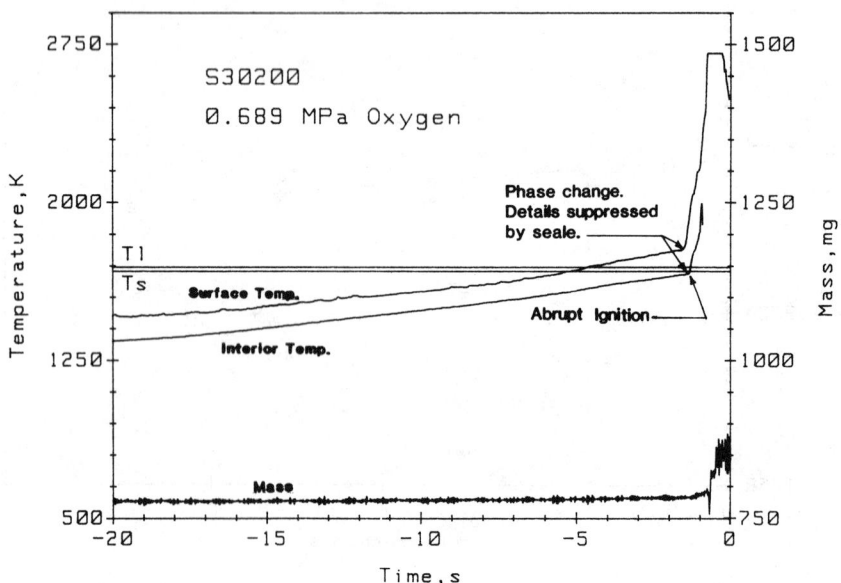

FIG. 8—*Experimental surface, interior and mass waveforms for S30200.*

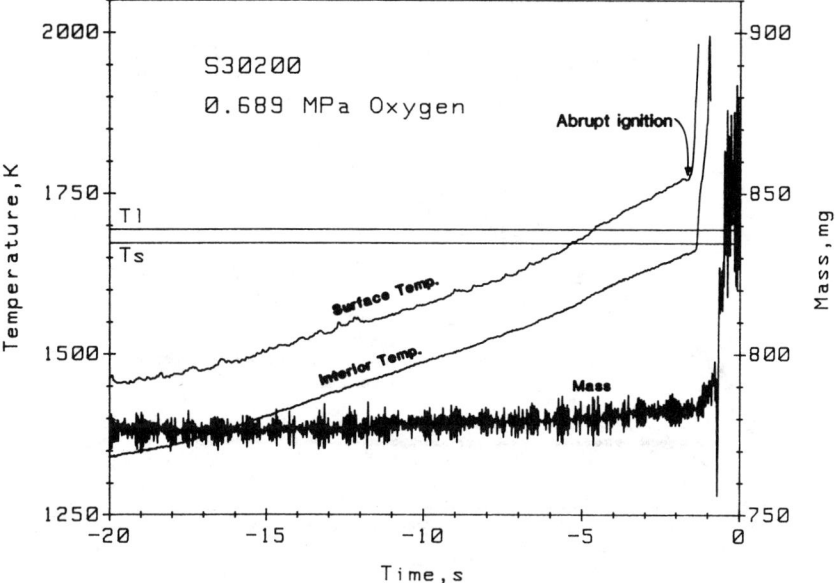

FIG. 9— *Waveforms of Fig. 8 expanded.*

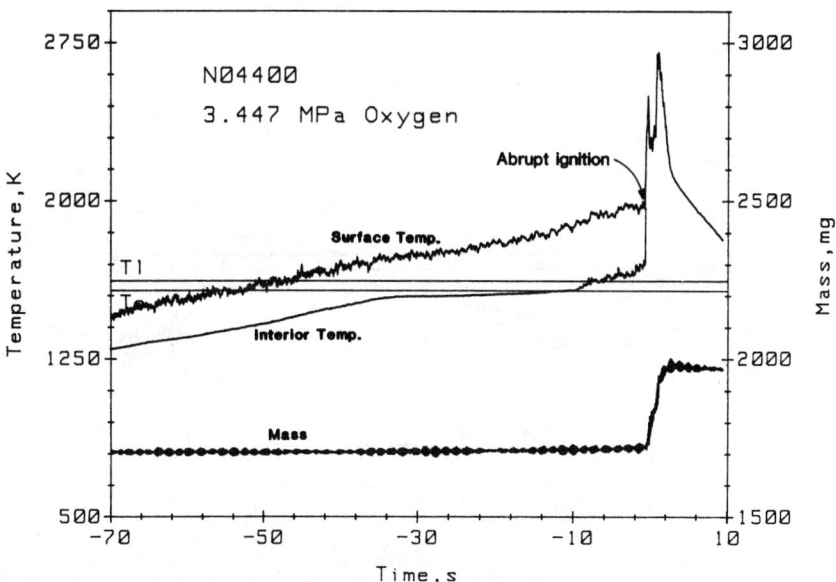

FIG. 10— *Experimental surface, interior and mass waveforms for N04400.*

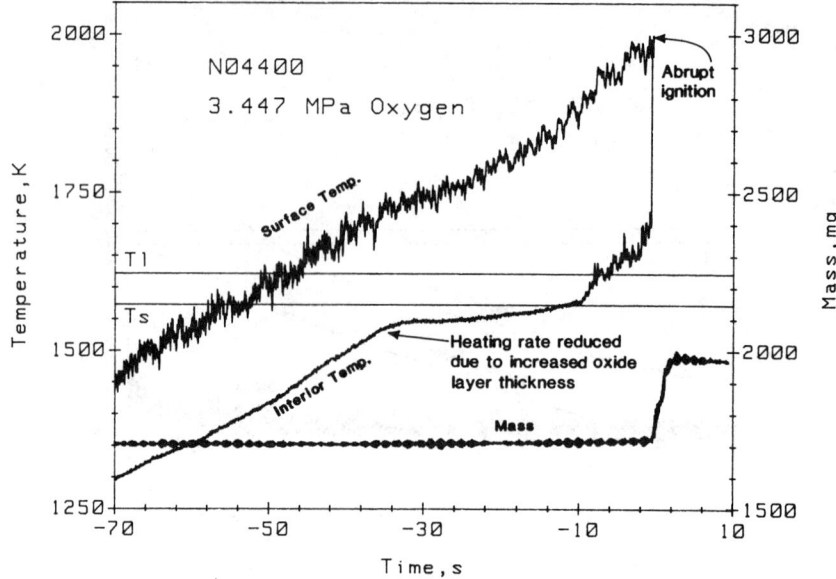

FIG. 11— *Waveforms of Fig. 10 expanded.*

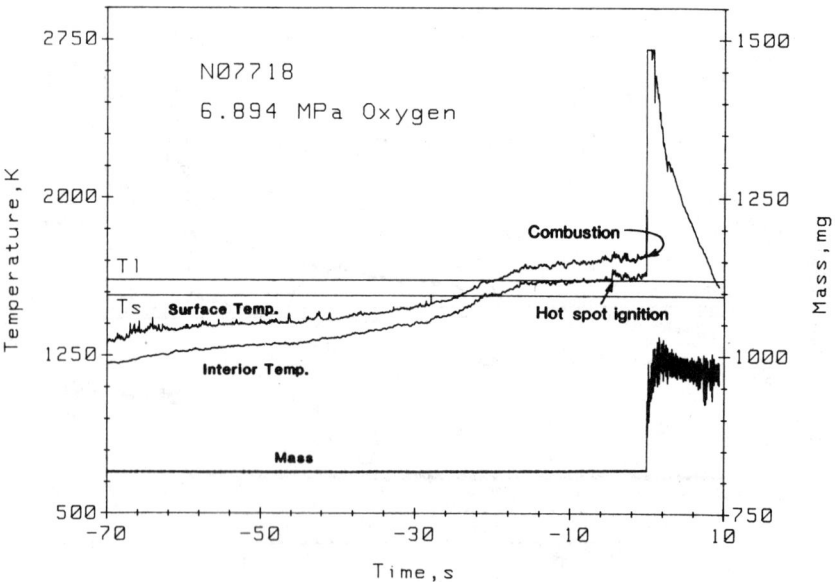

FIG. 12—*Experimental surface, interior, and mass waveforms for N-7718.*

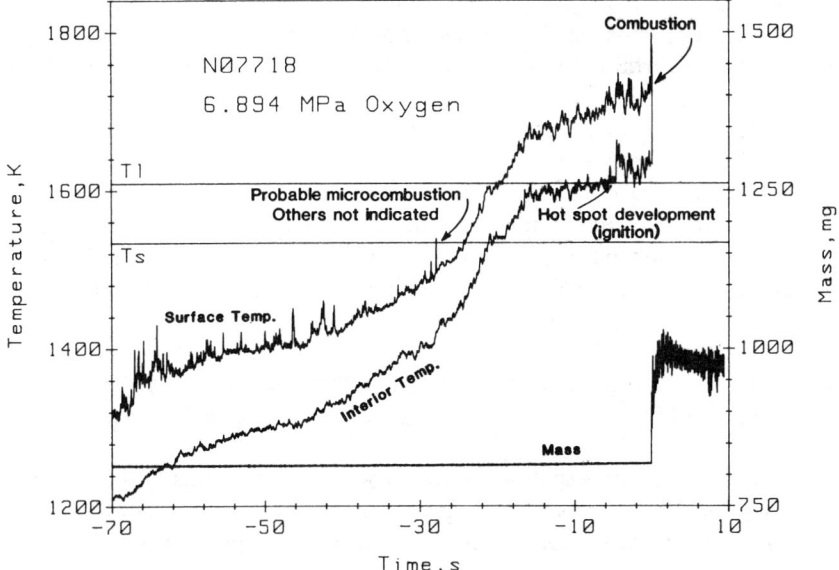

FIG. 13— *Waveforms of Fig. 12 expanded.*

The first and most obvious similarity between materials existed in the temperature waveforms. Figures 8, 10, and 12 show the variations that existed in the thermal activity for the temperature waveforms. There were three principal sources for this thermal activity.

The first source of thermal noise was refraction of the laser beam by convection currents generated by high temperatures at the specimen surface. This effect was most clearly evident in the irregular features of the surface temperature waveform for S30200 (Fig. 9). These convection currents caused the beam to drift randomly about the surface and to partially miss the specimen occasionally. Over a period of several seconds, this caused the surface temperature to vary from the general heating trend or even to cool when large deviations occurred. The magnitude of this effect was enhanced as oxygen pressure was increased.

The second source of thermal activity was the physical nature of the surface oxide layer. This layer had a physical structure that varied initially from a tightly bound protective layer to either a granular or a thin layered form at temperatures beyond approximately 1200 K. The laser beam generated high-temperature hot-spots on the layered structure because of the widely varying thermal conductivity of the surface. The net effect of these surface hot-spots was to produce a significant thermal noise background on the surface temperature waveform with occasional larger thermal spikes. N04400 and N07718,

Figs. 10 and 13, had a layered oxide surface and showed these effects. S30200 (Fig. 9) had a granular surface structure and did not, in general, show these effects. (Under certain conditions this surface may also develop hot-spots and generate thermal noise).

The third source of thermal activity appeared to be micro-combustion. A number of fast risetime thermal spikes were evident in the surface and interior temperature waveforms for certain materials; N07718 was one such material (Fig. 13). The maximum temperature rise detected in these thermal spikes varied widely but rarely exceeded 50 K. The amount of material consumed in these apparent micro-combustions was small and could not be detected because of effects of building vibration on the mass sensor. The risetime of the thermal spikes would suggest the sudden exposure of a reactive specie to the hot oxygen environment. The reactive specie(s) may be a minor component(s) of the material or may be generated by the oxidation process or both.

The second similarity that occurred between materials was in the ignition process. It should be stressed at this point that the ignition process was probably dependent upon the experimental setup to a large degree, at least for some materials. The experimental setup for which this discussion is considered valid was a top surface-heated specimen with an interior temperature gradient and an undisturbed oxide layer. Two ignition to combustion sequences have been observed, abrupt and progressive. A material may exhibit both sequences, often depending upon oxygen pressure. Three mechanisms appeared to be involved in the ignition process, phase changes, oxide decompositions, and micro-combustion. Examples of ignitions that involved phase changes and decompositions are those of S30200 and N04400 (Figs. 8 through 11).

The ignition mechanism for S30200 appeared to be the melting of ferrous oxide at 1650 K or the decomposition or melting of other iron oxides at temperatures greater than 1735 K. The ignition to combustion sequence could be abrupt (Fig. 9) or progressive (Fig. 14). The type of ignition to combustion sequence was probably a function of the energy released at the onset of ignition. The more energetic the reaction, the shorter the time between ignition and rapid combustion, since several oxide species had to be melted or decomposed before the onset of rapid combustion could begin. It should be pointed out that because of the axial thermal gradient it cannot be precisely determined whether many ignitions were due to the melting and subsequent oxidation of ferrous oxide or to the decomposition of ferric oxide. Ignition consistently occurred when the surface and interior temperatures were within the respective temperature ranges where these changes occur. However, the majority of the data appeared to indicate that the melting of ferrous oxide was the initiator of ignition.

For N04400 only an abrupt ignition to combustion sequence has been observed. The ignition mechanism appeared to be a major phase change in the

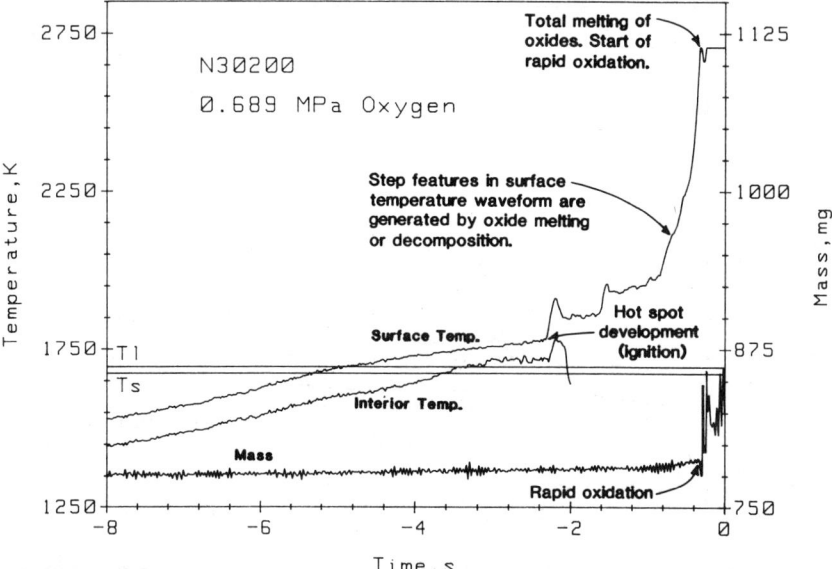

FIG. 14—*Expanded surface, interior, and mass waveforms for an S30200 ignition.*

oxide shell. This material oxidizes readily, especially in the liquid phase. For all practical purposes the specimen formed its own containment vessel; thus, this vessel must be melted or ruptured before rapid combustion can begin. This was achieved in these experiments when a major phase change(s) occurred in the oxide shell at surface temperatures greater than approximately 1950 K.

For N07718 the predominant ignition mechanism appeared to be micro-combustion. Both an abrupt and a progressive ignition-to-combustion sequence (Figs. 13 and 15) were produced. However, a complicating factor in the interpretation of the data was the fact that known phase changes and oxide decompositions took place in the temperature regions where some ignitions occurred. Thus, this material probably exhibited all ignition mechanisms to some extent. The ignition temperature was also a function of oxygen pressure.

The third similarity that occurred among surface burning materials was the combustion mechanism. For the experimental conditions employed in this study, rapid oxidation (combustion) could not occur until the oxide shell of the specimen was undergoing or had undergone failure by partial or complete melting. A clear demonstration of this fact is shown by the mass and surface temperature waveforms in Fig. 14. In fact, combustion could not occur if the liquid alloy was confined in a failproof container. Thus, the oxides produced by oxidation effectively prevented combustion under the appropriate condi-

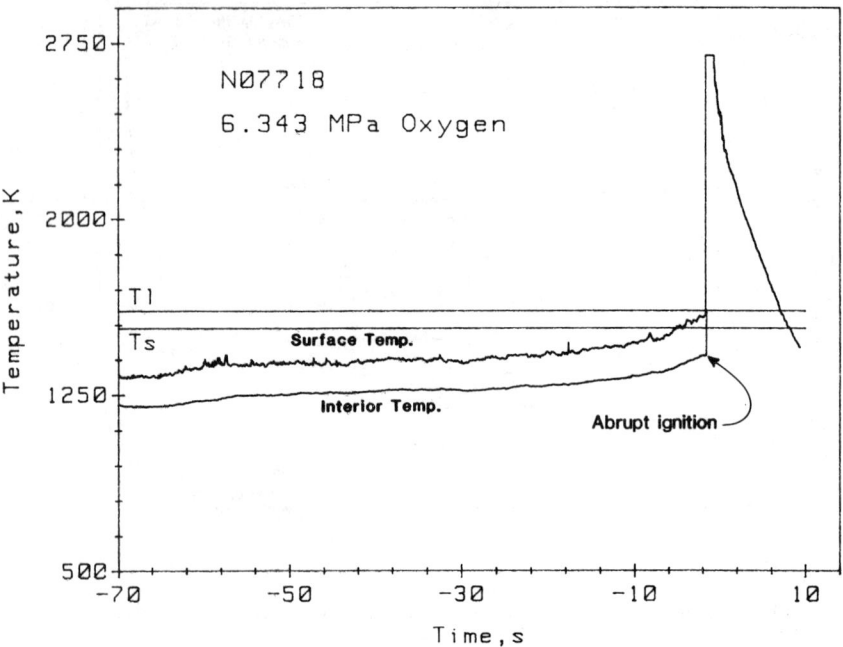

FIG. 15—*Typical abrupt ignition to combustion. sequence for N07718.*

tions even though there was evidence that the liquid oxides absorbed oxygen. Should external means be used to remove or break-up the oxide layer, the combustion temperature would probably be lowered from those observed in these experiments.

Figure 16 presents the ignition temperature versus oxygen pressure data for S30200. The ignition temperatures shown were derived from the interior thermocouple waveform; thus, the indicated temperature will be slightly less than the alloy-oxide interface temperature. As is seen, the ignition temperature was spread across the melting range at low oxygen pressures, but rapidly fell below the solidus temperature to the 1650 K temperature range, the melting point of ferrous oxide. It is, therefore, believed that most ignitions of this alloy were due to the ferrous oxide melting transition with other ignitions because of decompositions or phase changes in the oxide layer.

Figure 17 presents the ignition temperature data for N07718. These temperatures were also derived from the interior thermocouple waveform. The development of hot-spot ignitions or concurrent ignition/combustion appeared to occur as the result of microcombustion. Phase changes played a part in these microcombustions, but were not the sole cause. The ignition temperatures dropped below the solidus temperature at oxygen pressures of

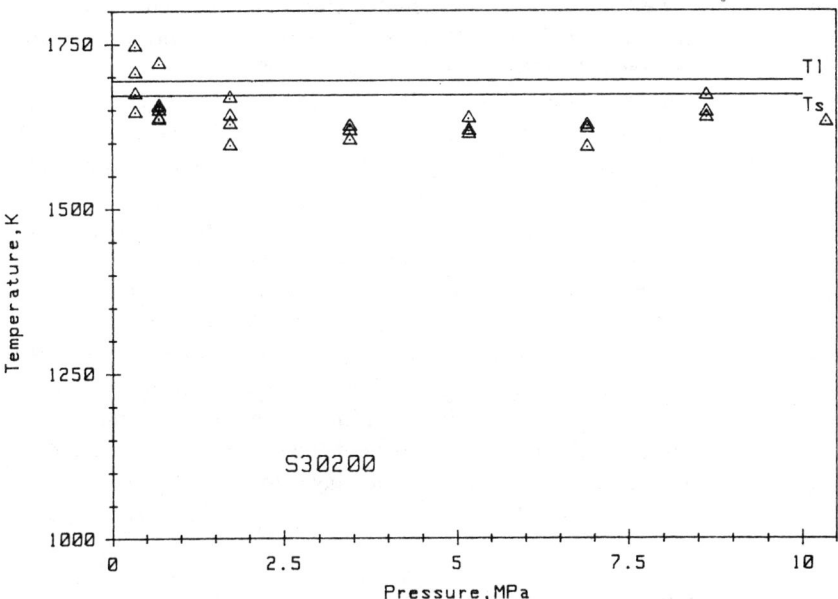

FIG. 16— *Variation of interior temperature at ignition with oxygen pressure for S30200.*

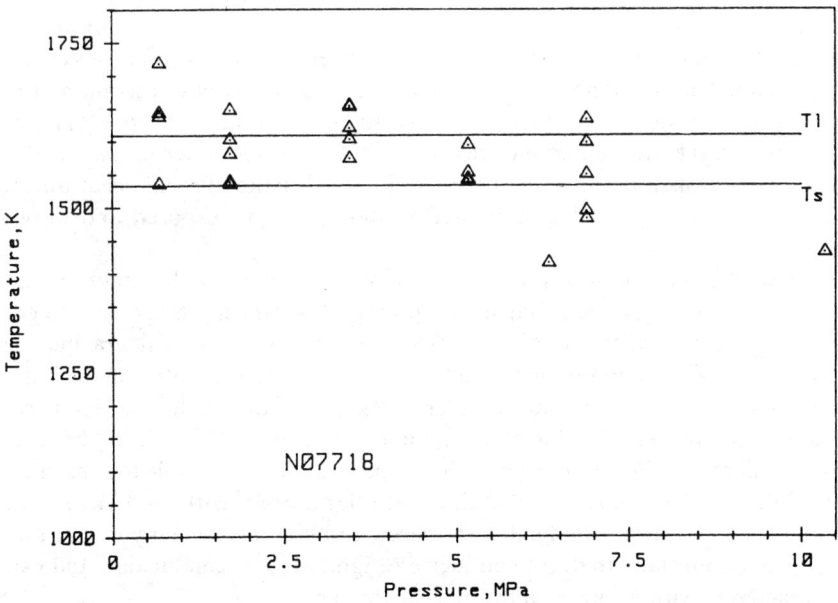

FIG. 17— *Variation of interior temperature at ignition with oxygen pressure for N07718.*

6.894 MPa and beyond. They are expected to continue to drop because of the existence of microcombustions and the existence of one or more volatile species at temperatures below the present ignition temperatures. These volatile species may ultimately provide an ignition source at higher oxygen pressures.

An ignition temperature versus pressure curve for N04400 is not shown because of experimental difficulties in obtaining interior temperature data. This material, as it liquifies, readily penetrated the thermocouple insulation, alloyed with the thermocouple, and destroyed the data beyond the solidus temperature. From the behavior of the limited thermocouple data that exists, this material did not have an ignition temperature in the sense that S30200 and N07718 had. The material developed oxidation rates consistent with combustion at a major phase change in the oxide layer, at approximately 1950 K, at all oxygen pressures tested to date. However, it was considered that the material could be induced into rapid oxidation at the liquidus temperature, and probably lower, with the rapid removal of the oxide layer because of the rapid oxidation that occurred with the oxide layer intact.

Conclusions and Recommendations

It was clear from the temperature and mass data, which were basically thermograms and oxidation rate data, that alloy ignition was a fairly complex process. The use of an oxygen diffusion model as the ignition initiation mechanism does not appear appropriate. For surface burning alloys the ignition mechanisms appeared to be the sudden exposure of sufficient quantities of reactive material to the hot oxygen environment, decomposition of certain oxides and phase changes. For vapor burning alloys, in our case aluminum alloys, the ignition mechanism appeared to be the contact of molten aluminum with a reactive material or restrictive heat transfer under normal oxidative or oxide shell failure conditions. Major combustion did not occur unless the materials were fluid and allowed to move, that is, exposed unoxidized surface.

To define the conditions for ignition more precisely, the thermogram and oxidation rate data produced in the present experiments must be refined. This would entail differential thermal analysis (DTA) and thermogravimetric analysis (TGA) studies under the appropriate conditions. Unfortunately, the equipment to do these studies under the appropriate conditions, temperatures to greater than 2300 K and oxygen pressures to 70 MPa, do not exist. It is considered by the author that such equipment may be possible to construct. To determine if this is true, feasibility studies are presently underway. The present laser combustion facility, however, can identify the temperature and pressure conditions that are conducive to ignition and combustion and can characterize with accuracy most metallic materials.

Acknowledgments

The author would like to express his gratitude to Mr. Phillip Billiard, Mr. James D. Breuel, Mr. James A. Hurley, and Mr. Ke Nguyen for their assistance in this study, and to the George C. Marshall Space Flight Center, Mr. John G. Austin, Jr., technical representative, for supporting this program.

Kenneth McIlroy[1] and Robert Zawierucha[1]

The Use of the Accelerating Rate Calorimeter in Oxygen Compatibility Testing

REFERENCE: McIlroy, K. and Zawierucha, R., **"The Use of the Accelerating Rate Calorimeter in Oxygen Compatibility Testing,"** *Flammability and Sensitivity of Materials in Oxygen-Enriched Atmospheres: Second Volume, ASTM STP 910*, M. A. Benning, Ed., American Society for Testing and Materials, Philadelphia, 1986, pp. 98–107.

ABSTRACT: A number of different test methods have been utilized to assess the compatibility of various materials or substances with oxygen. Accelerating rate calorimetry is a relatively new analytical technique that has been used to detect adiabatic, self-heating reactions at rates as slow as 0.03°C/min. In this study six materials were tested by the accelerating rate calorimeter and the conventional autoignition apparatus in oxygen over a wide range of pressures.

The results showed that exothermic reactions can occur well before the autogenous ignition temperature (AIT) for some materials. For example, nitrile rubber demonstrated that the onset of adiabatic self-heating occurred about 40 to 53°C below the AIT, and "MOBIL" DTE 25 compressor oil showed a range of 38 to 73°C below AIT in the 13.6- to 36.7-MPa (2000- to 5400-psi) pressure range.

KEY WORDS: oxygen, calorimetry, oxygen compatibility, autogenous ignition temperature, accelerating rate calorimeter, exothermic reactions, nonmetallics

On a traditional basis oxygen suppliers and users individually developed their own methods to test and select engineering materials for oxygen systems. ASTM Committee G-4 on Compatibility and Sensitivity of Materials in Oxygen Enriched Atmospheres was formed in 1975 to achieve a consensus view that would standardize tests by oxygen suppliers, oxygen users, and vendors of equipment for the oxygen industry.

To a large extent the consensus has been achieved by the development of a number of standard tests for assessing the suitability of materials for oxygen applications.

[1]Consultant and manager, respectively, Union Carbide Corp., Linde Division, P.O. Box 44, Tonawanda, NY 14150.

Typical industry tests and the applicable ASTM standard include the following:

- ASTM Test Method for Autogenous Ignition Temperature of Liquids and Solids in a High-Pressure Oxygen-Enriched Environment (G 72)
- ASTM Methods for Heating of Combustion of Liquid Hydrocarbon Fuels by Calorimeter (D 240)
- ASTM Test Method for Compatibility of Materials with Liquid Oxygen (Impact Sensitivity Threshold and Pass-Fail Technique) (D 2512)
- ASTM Test Method for Ignition Sensitivity of Materials to Gaseous Fluid Impact (G 74)
- ASTM Method for Measuring the Minimum Oxygen Concentration to Support Candle-Like Combustion of Plastics (Oxygen Index) (D 2863)
- ASTM Test Method to Determine the Ignition Sensitivity of Materials to Mechanical Impact in Pressurized Oxygen Environments (G 86)
- Promoted Ignition (no current standard)
- Service simulation (application dependent)

It must be recognized that current procedures are not the last word. New test procedures may be required for specific applications involving oxygen. New technologies may also complement current test procedures. It is important for the industry to stay abreast of new technologies and techniques for assessing the compatibility of materials in oxygen.

The accelerating rate calorimeter is a relatively recent development, which has been used to evaluate thermal and pressure hazard potential. In contrast to a conventional autoignition test it is able to detect adiabatic self-heating below the autoignition temperature. Severe self-heating could lead to a thermal runaway and ignition. The following data may be generated in an experiment:

- Heat generation rates
- Adiabatic reaction temperature
- Adiabatic reaction pressure
- Adiabatic self-heating parameters
- Temperature versus real time
- Time to maximum rate of reaction
- Pressure rate data

At the present time a test method for the accelerating rate calorimeter is under consideration by ASTM Committee E-27 on Hazard Potential of Chemicals. However, this method does not address testing in oxygen atmospheres.

It is the purpose of this paper to discuss results obtained via the accelerating rate calorimeter. A comparison of the data will be made with the conventional autoignition test (ASTM G 72).

Experimental Procedure

Autoignition Tests

These tests were conducted in pure oxygen over a nominal pressure range of 13.6 to 36.7 MPa (2000 to 5400 psi). Up to a test pressure of 13.6 MPa (2000 psi) the apparatus was essentially identical to that described in ASTM Procedure G 72. An Inconel bomb was substituted for the 316 stainless steel for tests above 13.6 MPa (2000 psi).

Accelerating Rate Calorimeter Tests

The accelerating rate calorimeter tests were conducted in pure oxygen over the same pressure range as the autoignition tests. The apparatus shown in Fig. 1 is considerably different from a conventional autoignition experiment. The equipment consists of a calorimeter, which contains the bomb, a blast protective clam shell to contain the test, a computer to monitor and control the heating, an *X-Y* plotter, and an auxiliary strip chart recorder.

Figure 2 shows the "Hastelloy C-276" bombs used in the oxygen environments. The heavy bomb was used in all tests. Bombs fabricated from other

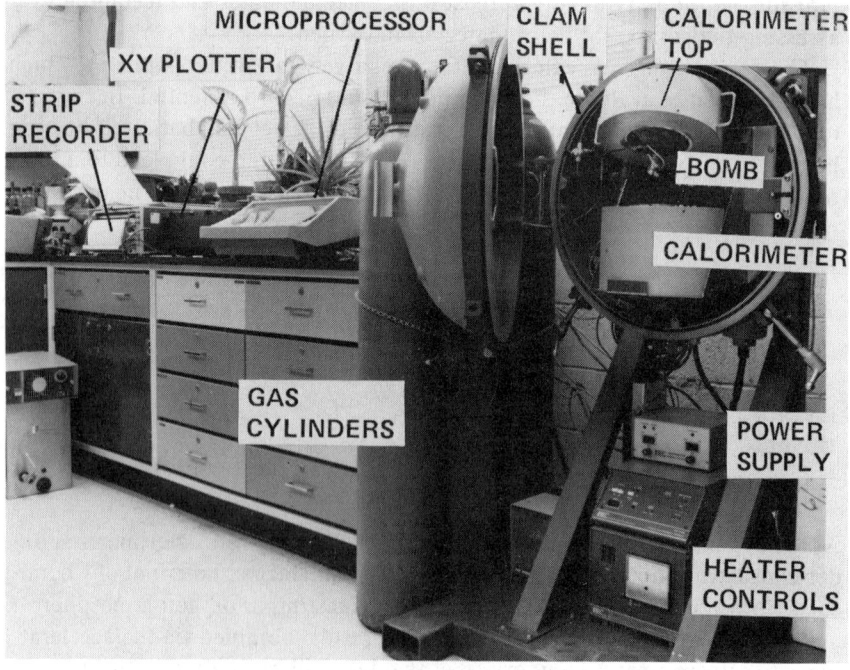

FIG. 1—*Accelerating rate calorimeter (ARC®, Columbia Scientific Corp., Austin, TX).*

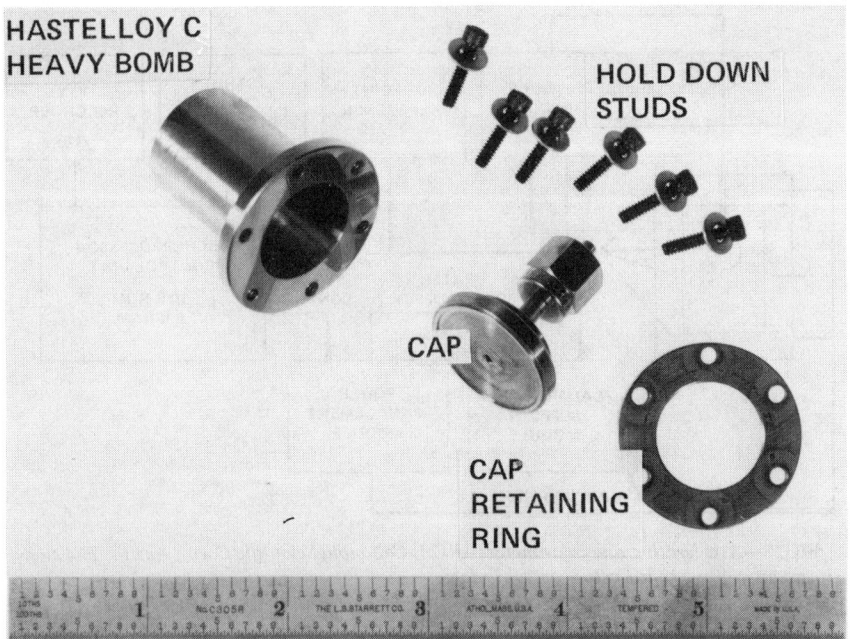

FIG. 2—*Photograph of ARC® (Columbia Scientific Corp., Austin, TX) bomb.*

alloys are available but may not be suitable for oxygen compatibility testing. In contrast to the bombs used in conventional autoignition tests these are relatively small, and the test samples are also relatively small being on the order of 0.1 g. Figure 3 shows the accelerating rate calorimeter analysis system and microprocessor control system.

Typically, the material was loaded into the test bomb and slowly pressurized with oxygen to avoid adiabatic heating. Heating steps were used to elevate the sample temperature followed by wait and search periods to monitor for self-heating reactions indicative of exotherms. Once an exotherm was detected above a predetermined threshold limit of 0.03°C/min, the system allowed the exotherm to go to completion but maintained the bomb at adiabatic conditions. All data were recorded and plotted on the X-Y recorder in a heat rate versus temperature format. It should be noted that depending on the instructions programmed into the device a test may take many hours to complete. The automated features of the test are therefore very useful.

Materials

The materials selected for the program were chosen to provide a wide response to the oxygen environment and therefore allow a comparison of the

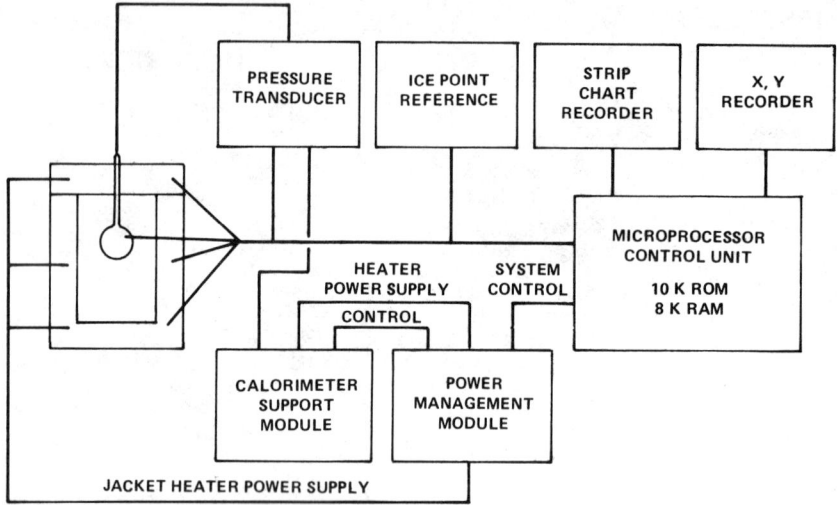

FIG. 3—*Accelerating rate calorimeter (ARC®, Columbia Scientific Corp., Austin, TX) analysis system and microprocessor control system.*

two test methods. Table 1 shows the test materials, their generic classification, and their heats of combustion.

Results and Observations

Autoignition Tests

Table 2 is a summary of the results of the autoignition tests. "Mobil" DTE-25 was tested over the pressure range of 1.0 to 36.7 MPa (150 to 5400 psi) to show the effects of increasing oxygen pressure on a hydrocarbon oil. The greatest effects on the autogenous ignition temperature were at the lower pressures. There was little change in the autogenous ignition temperature

TABLE 1—*Heats of combustion for test materials.*

Materials	Manufacturer	Description	Heat of Combustion, cal/g
DTE-25	Mobil	hydrocarbon oil	10 700
Teflon® TFE	DuPont	polytetrafluoroethylene	1 700
Nylon 101	DuPont	Type 6/6 Nylon	7 500
KEL-F 81	3M Co.	Trifluorochloroethylene KF-6060	2 300
Buna N	Reeves Brothers	Nitrile rubber, Style 7478	8 000
MRC Compound 514 AD	Minnesota Rubber Co.	FKM fluoroelastomer	4 200

TABLE 2—*Autogenous ignition temperature test results.*

Material	Nominal, MPa	Pressure, psi	Autogenous Ignition Temperature, °C
DTE-25	1.0	150	300
	3.7	550	270
	13.6	2000	220
	36.7	5400	224
Teflon® TFE	13.6	2000	>400
	36.7	5400	>400
Nylon 101	13.6	2000	205
	36.7	5400	209
KEL-F 81	13.6	2000	>400
	36.7	5400	>400
Buna N	13.6	2000	195
	36.7	5400	185
MRC Compound 514 AD	13.6	2000	320
	25.1	3700	290
	36.7	5400	250

when the pressure was increased from 13.6 to 36.7 MPa (2000 to 5400 psi).

The shifts in the autogenous ignition temperature of materials, such as Teflon® TFE, Nylon 101, "KEL-F" 81, and Buna N with increasing oxygen pressure in the 13.6 to 36.7 MPa (2000 to 5400 psi) range were considered to be insignificant.

However, the Minnesota Rubber Company (MRC) Compound 514 AD did show an oxygen pressure effect in its autogenous ignition temperature. It dropped 70°C when the oxygen pressure was increased from 13.6 to 36.7 MPa (2000 to 5400 psi). An intermediate pressure test at 25.1 MPa (3700 psi) shown in Table 2 confirmed the pressure dependency.

Accelerating Rate Calorimeter Tests

Table 3 contains the results of the accelerating rate calorimeter tests. The materials tested in this phase are from the same lot as that reported on in Table 2.

The data reported in Table 3 require some explanation. The column entitled "Onset of Exotherm" represents that minimum temperature at which a self-heating rate of at least 0.03°C/min was detected. This is a relatively slow heating rate, but it is indicative of an interaction between the material and the oxygen environment.

The column entitled "Temperature of Peak Reaction" is self explanatory. Data in this column would be expected to be in relatively close proximity to the autogenous ignition temperature. It should be noted that a considerable

TABLE 3—Accelerating rate calorimeter results.

Material	Nominal, MPa	Pressure, psi	Onset[a] of Exotherm, °C	Temperature[b] of Peak Reaction °C	Autogenous[c] Ignition Temperature (AIT) °C	Comment
Mobil DTE-25	1.8	270	183	280		
	3.3	480	184	296		
	13.6	2000	182	206	220	
	36.7	5400	151	193	224	
Teflon® TFE	13.6	2000	393	394	>400	
	36.7	5400	373	374	>400	sample intact at 400°C
Nylon 101	13.6	2000	187	191	205	
	36.7	5400	165	210	209	pressure spike stopped test
KEL-F 81	13.6	2000	361	362	>400	
	36.7	5400	none detected	372	>400	pressure spike stopped test
Buna N	13.6	2000	142	210	195	
	36.7	5400	145	186	185	
MRC Compound 514 AD	13.6	2000	287	289	320	
	36.7	5400	195	230	250	

[a]Detection of self heating at a rate >0.03°C/minute.
[b]Peak Reaction is either peak heating rate or pressure spike.
[c]Data from Table 2 shown for direct comparison.

amount of time could elapse between the onset of an exotherm and the temperature of maximum heating rate or a pressure spike.

The DTE-25 oil was tested in the pressure range of 1.8 to 36.7 MPa (270 to 5400 psi). Note that the onset of an exotherm was independent of pressure up to 13.6 MPa (2000 psi) and occurred within a narrow temperature range of 182 to 184°C. Above that pressure the temperature at which the maximum heating rate was achieved was highly pressure dependent and compared closely with the autogenous ignition temperature data in Table 2.

Discussion

From the standpoint of McQuaid et al [1] and the work of Nihart and Smith [2] one would expect the following:

1. The effect of oxygen pressure on a hydrocarbon and its autogenous ignition temperature would be most pronounced at lower oxygen pressures.

2. The difference in the autogenous ignition temperature of nonmetallic engineering materials caused by an increase in oxygen pressure from 13.6 to 36.7 MPa (2000 to 5400 psi) would be slight.

The current test results confirm these expectations with one exception. The material identified as MRC Compound 514 AD showed a substantial decrease in its autogenous ignition temperature as a result of an increase in oxygen pressure from 13.6 to 36.7 MPa (2000 to 5400 psi). This fact demonstrated that oxygen pressure may be significant in the behavior of some materials in the high-pressure regime, and the necessity for high-pressure tests should not be discounted. The increased reactivity of the MRC Compound 514 AD with increasing pressure was corroborated by another technique, the accelerating rate calorimeter.

Significant differences between the Accelerating Rate Calorimeter test and a conventional autogenous ignition temperature test include the following:

(1) improved temperature measurement capability in the accelerating rate calorimeter,

(2) smaller sample sizes,

(3) a longer test duration in which the specimen is able to interact with the oxygen environment, and

(4) slower heating rates.

The output of the accelerating rate calorimeter that is most comparable to the autogenous ignition temperature is the temperature at which the maximum heating rate or a pressure spike was recorded. A comparison of data in Tables 2 and 3 showed that this temperature is somewhat lower than an autogenous ignition temperature determined by ASTM G 72, but it still is relatively close. It might be hypothesized that the longer duration of the accelerating rate calorimeter test at elevated temperature might result in material changes caused by interaction with oxygen. This might be verified by autoignition tests of deliberately aged materials.

As to the interaction of nonmetallic materials with oxygen at temperatures below their autoignition point there is little doubt, as shown in Table 3, that the accelerating rate calorimeter effect picked this up. With the exception of "KEL-F" 81 all of the test materials exhibited the onset of an exotherm indicative of a minimum self-heating rate greater than 0.03°C/min. At the higher pressure level "KEL-F" 81 was a notable exception. It exhibited no self-heating reaction until it spontaneously reacted with the oxygen, was consumed, and produced a pressure spike. It should be noted that "KEL-F" 81 has a melting point of about 212°C, well below the reaction temperature.

The significance of self-heating may vary. Some materials self-heat at a slow rate that can be handled easily. Other self-heating reactions may feed upon themselves and reach a temperature of no-return followed by a thermal runaway. The existence of self-heating reactions, however slight, may also be an indication of a long-term aging reaction, which could result in a deterioration of physical or mechanical properties. In this regard, the accelerating rate calorimeter offers potential as a device for material development to assure stability in service and material selection in critical environments.

Currently ASTM Guide for Evaluating Nonmetallic Materials for Oxygen Service (G 63) suggests that the autogenous ignition temperature (measured at use or greater pressure) of a nonmetallic material be at least 100°C greater than its service temperature. The 100°C margin is empirical to compensate for variations in real world versus laboratory conditions and variations in normal operations.

At high pressures this appears to be a conservative approach as the accelerating rate calorimeter did not detect self-heating reactions more than 100°C below the autogenous ignition temperature for the tested materials. At lower pressures (3.3 MPa), however, self-heating of the "Mobil" DTE-25 hydrocarbon oil was observed at a temperature greater than 100°C below the autogenous ignition temperature.

The accelerating rate calorimeter is useful in critical applications where it is suspected that the 100°C margin, as measured in the AIT apparatus, does not really exist, that is, reactions start to occur at lower temperatures and the peak reaction does not occur until the AIT is attained by the sample. The long-term effect on the properties of nonmetallic materials by thermal aging is of concern in oxidizing environments, and the accelerating rate calorimeter may have potential in this area. Thermal aging may be accompanied by self-heating.

Summary

While the general trend for nonmetallic materials may be to show little difference between the autogenous ignition temperature at 13.6 MPa (2000 psi) and more elevated pressures, this is by no means a certainty. One of the six materials in the program showed a pronounced dependence of autogenous ignition temperature on oxygen pressure.

The temperatures at which a pressure spike occurred or the accelerating rate calorimeter measured a maximum heating rate were reasonably close to the autogenous ignition temperatures measured by ASTM G 72. The self-heating behavior of the tested materials varied considerably. One material, "KEL-F" 81, exhibited no self-heating before it spontaneously ignited at the elevated pressure.

The ASTM G 63 suggestion that the autogenous ignition temperature of nonmetallic material be at least 100°C above the service temperature appears to be reasonably conservative, at high pressures. At high pressure none of the tested materials exhibited self-heating more than 100°C below the autogenous ignition temperature.

The accelerating rate calorimeter is expected to be a valuable tool for characterizing materials in oxygen environments and will supplement other test methods. It is more sensitive to the detection of adiabatic self-heating than the conventional autoignition apparatus. This feature indicates that it can be used to more accurately characterize material: environment interactions that may lead to degradation in long-term service or assess the effects on long-term degradation of oxygen compatibility.

Acknowledgments

The contributions of Messrs. R. W. Zimmerman and R. F. Alessi in the conducting of the test program are acknowledged.

References

[1] McQuaid, R. W., Sheets, D. G., and Bieberich, J., "Determination of Autogenous Ignition Temperatures of a Steam Turbine Lubricating Oil in Nitrogen and Oxygen Mixtures," *Flammability and Sensitivity of Materials in Oxygen-Enriched Atmospheres, STP 812*, B. L. Werley, Ed., American Society for Testing and Materials, Philadelphia, 1983, pp. 43–55.
[2] Nihart, G. J. and Smith, C. P., "Compatibility of Materials with 7500 PSI Oxygen," DDC AD 608260, AMRL-TDR-64-76, Union Carbide Corporation, Linde Division, Tonawanda, NY, Oct. 1964.

Coleman J. Bryan[1] *and Robert Lowrie*[2]

Comparative Results of Autogenous Ignition Temperature Measurements by ASTM G 72 and Pressurized Scanning Calorimetry in Gaseous Oxygen*

REFERENCE: Bryan, C. J. and Lowrie, R., **"Comparative Results of Autogenous Ignition Temperature Measurements by ASTM G 72 and Pressurized Scanning Calorimetry in Gaseous Oxygen,"** *Flammability and Sensitivity of Materials in Oxygen-Enriched Atmospheres: Second Volume, ASTM STP 910*, M. A. Benning, Ed., American Society for Testing and Materials, Philadelphia, 1986, pp. 108–117.

ABSTRACT: The autogenous ignition temperature of four materials was determined by ASTM (G 72) and pressurized differential scanning calorimetry at 0.68-, 3.4-, and 6.8-MPa oxygen pressure. All four materials were found to ignite at lower temperatures in the ASTM method. The four materials evaluated in this program were Neoprene®, Vespel SP-21®, Fluorel E-2160®, and nylon 6/6.

KEY WORDS: heat measurement, oxygen, pressure, autogenous ignition temperature, Neoprene®, Vespel SP-21®, Fluorel E-2160®, nylon 6/6

The temperature at which a material will spontaneously ignite without the application of a flame or spark is called the autogenous ignition temperature (AIT). The AIT has been found to vary inversely with the oxygen pressure or the oxygen partial pressure in mixed gases. This temperature variation has been shown to approach a lower limit in the 6.8 to 10.3 MPa (1000 to 1500 psi) range and then to remain essentially constant to at least 51 MPa (7500 psi) [1].

---

*ASTM Test Method for Autogenous Ignition Temperature of Liquids and Solids in a High-Pressure Oxygen-Enriched Environment (G 72).

[1] Aerospace technology materials engineer, John F. Kennedy Space Center, NASA, Kennedy Space Center, FL 32899.

[2] Senior research metallurgist, The BOC Group, Inc., Technical Center, Murray Hill, NJ 07874.

During the early 1970s, the David Taylor Naval Ship R&D Center in An-napolis, MD, developed a pressurized AIT test apparatus and procedure. This procedure was subsequently proposed to ASTM Committee G-4 on Compatibility and Sensitivity of Materials in Oxygen Enriched Atmospheres as a candidate for development into an ASTM test method. The method was approved and published as ASTM Test Method for Autogenous Ignition Temperature of Liquids and Solids in a High-Pressure Oxygen-Enriched Environment (G 72). In this method, a 0.2- to 0.5-g sample is placed in a glass sample holder assembly that is then placed in a high-pressure reaction vessel. After pressurization to the test pressure, the vessel is then heated at 5°C/min until ignition is detected. One problem encountered with this procedure is maintaining a uniform heating rate at higher temperatures.

With the development of a pressurized differential scanning calorimeter (PDSC) by the DuPont Instruments Division, it became possible to perform AIT measurements utilizing much smaller samples and with a much more precisely controlled heating rate. In addition, the computerized controller permits storage, playback, and detailed analysis of the test data. To increase the accuracy of the pressure measurement, a 0- to 13.6-MPa (2000-psi) pressure transducer was added to the PDSC bleed port.

The purpose of this program was to determine the AIT of several selected materials using the ASTM G 72 and PDSC apparatus and compare the results.

Experimental Methods and Materials

The ASTM G 72 AIT's were determined at the BOC Group, Inc., Technical Center, Murray Hill, NJ. The apparatus and procedure were as described in the ASTM test method.

The PDSC tests were performed at the John F. Kennedy Space Center, FL, using a commercially available PDSC system. This system consisted of a digital programmer with plotter, a dual disk drive data recorder, a differential scanning calorimeter cell base, a PDSC test cell rated for use at pressures up to 6.8 MPa (1000 psi), and a standard data analysis software program. The PDSC test cell is shown in Fig. 1.

The minimum temperature required to cause the sample to ignite spontaneously was determined at 0.68, 3.4, and 6.8 MPa (100, 500, and 1000 psi). The temperature at which spontaneous ignition occurred is denoted by a sudden increase of heat flow in the sample.

Initial experiments were conducted to determine the effect of sample size, sample configuration, and heating rate. Several tests were run over the range of 0.5 to 16 mg to study the effect of sample size. The configuration study varied the sample from a single piece to as many as ten small pieces for a given test. The effect of heating rate was studied at 5, 10, and 50°C/min.

For PDSC AIT determinations, the sample was first cut into seven to ten

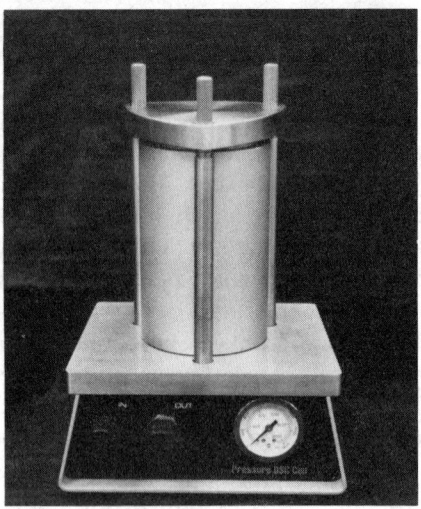

FIG. 1—*PDSC test cell, rated for use to 6.8 MPa and 600°C in an oxidating environment.*

small pieces, weighed in a tared, open platinum sample cup, and placed in the PDSC cell as shown in Fig. 2. Then, the cell was closed and purged with oxygen for 5 min at 10 to 50 mL/min. Next, the cell was pressurized twice to the approximate test pressure and released. The cell was then pressurized to slightly above the desired test pressure, allowed to stand for 10 min while the

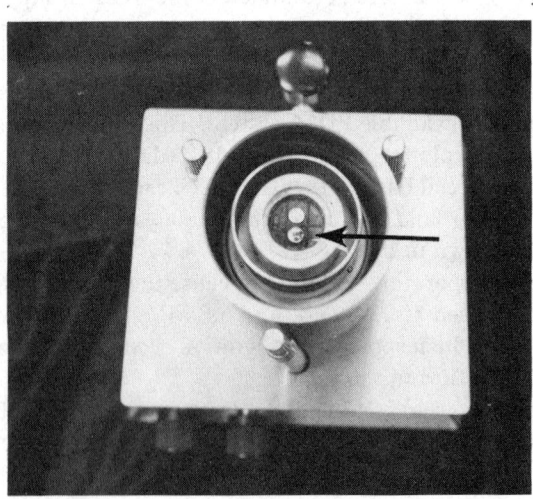

FIG. 2—*PDSC test cell depicting the sample chamber. The arrow points to the sample cup containing the cut samples.*

gas cooled, and the pressure adjusted to the final test pressure. The cell was rapidly heated to 50°C where a programmed heating rate of 10°C/min was begun. The cell was programmed to heat to 500°C; however, the test was usually terminated manually approximately 50°C above the indicated AIT.

The materials chosen for this study were Neoprene®, Vespel SP-21®, Fluorel E-2160®, and nylon 6/6.

Results and Discussion

Initial tests conducted with the PDSC using Neoprene® at 3.4 MPa revealed essentially no variation in AIT at heating rates of 5, 10, and 50°C/min. Therefore, a heating rate of 10°C/min was selected for all subsequent tests.

The next series of tests was conducted to study the variation of AIT with particle size. It was found that large pieces of Neoprene® exhibited a higher AIT than did several small pieces. By varying the number of pieces between one and ten for an approximate 15-mg sample, it was found that all tests using four or more pieces gave essentially the same AIT. Finely ground powders were not evaluated.

The third experimental variable identified was the sample weight. Ignition and combustion of polymeric materials usually occur in the gas phase. In order for ignition to occur, a flammable mixture must first exist. This mixture is developed by thermal decomposition and outgassing of the material. Obviously, the larger the sample, the faster the concentration of combustible gases should increase in a confined space until some limiting size is reached. Therefore, it would be expected that the AIT would decrease with increasing sample weight to some constant value. This limiting weight would primarily be a function of the enclosed volume and the decomposition and outgassing rates of the material.

Except for Neoprene®, the AIT was found to decrease to a limiting value with increasing sample weight as shown in Table 1. However, the Neoprene® AIT was observed to drop suddenly between 12.0 and 12.9 mg of sample weight. Examination of the PDSC thermogram revealed that the exothermic reaction peak, occurring at approximately 220°C (Fig. 3), gradually increased with increasing sample weight. In the 12-mg thermogram (Fig. 4), this peak exhibited a very small loop at the tip that indicated an additional exothermic event had occurred. This event may have been a flash of the decomposition products. Finally, ignition occurred with a 12.9-mg sample in this lower temperature range (Fig. 5). Therefore, a sample weight range from 14 to 16 mg was selected for all materials.

Once the experimental variables were determined, the AIT's of the four test materials were measured in triplicate at 0.68, 3.4, and 6.8 MPa. These results along with the ASTM G 72 AIT's are summarized in Table 2.

It is interesting to note that even though the PDSC AIT's are all higher than those determined by the ASTM method, the ranking of the materials is

TABLE 1—*Effect of sample weight on AIT by PDSC.*

| | Pressure, MPa | | | |
| | 3.4 MPa | | 6.8 MPa | |
Material	Weight, mg	AIT, K	Weight, mg	AIT, K
Vespel SP-21®	3.1	721	3.0	692
	6.0	724	6.2	687
	8.9	711	9.0	679
	12.1	700	12.2	668
	15.6	694	15.3	671
Nylon 6/6	...	...	3.1	NI[a]
	...	...	6.1	626
	...	...	9.0	604
	...	...	11.8	544
	...	...	15.2	543
Neoprene®	0.5	679	...	...
	1.5	674	...	...
	3.5	653	...	...
	7.0	643	...	...
	8.0	652	...	...
	9.0	640	...	...
	12.0	632	...	...
	12.9	482	...	...
	14.7	476	...	...

[a]NI = no ignition.

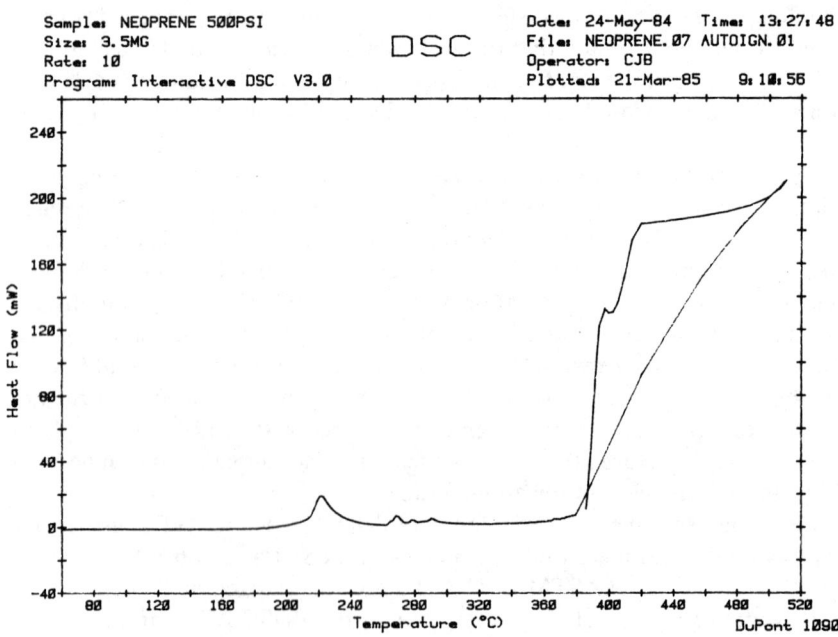

FIG. 3—*Thermogram for 3.5-mg of Neoprene® at 4.5 MPa. Ignition occurred at 380.3°C. Note exothermic reaction peak at 220°C.*

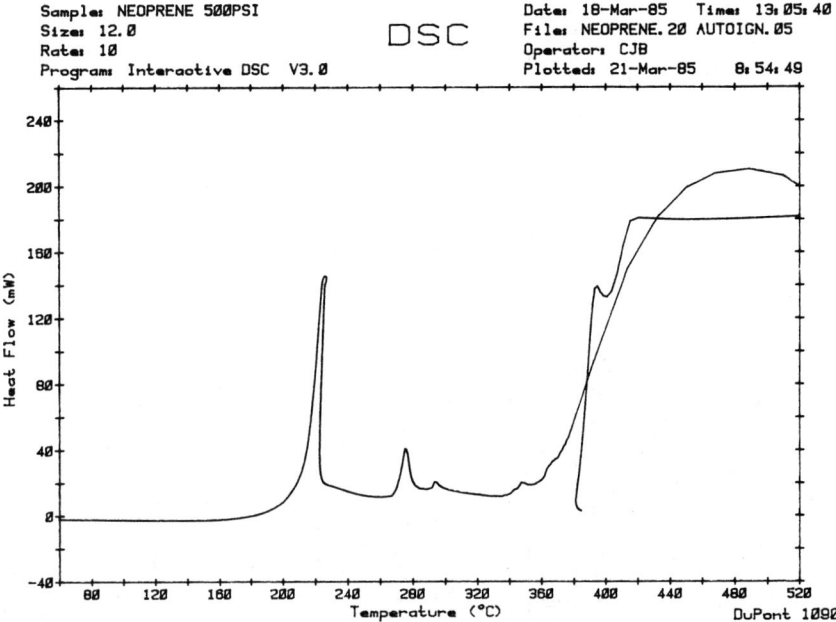

FIG. 4—*Thermogram for 12.0-mg sample of Neoprene® at 3.4 MPa. Ignition occurred at approximately 375°C. Note the loop at the tip of the exothermic reaction peak at 220°C. This loop may be indicative of a flash of the decomposition products.*

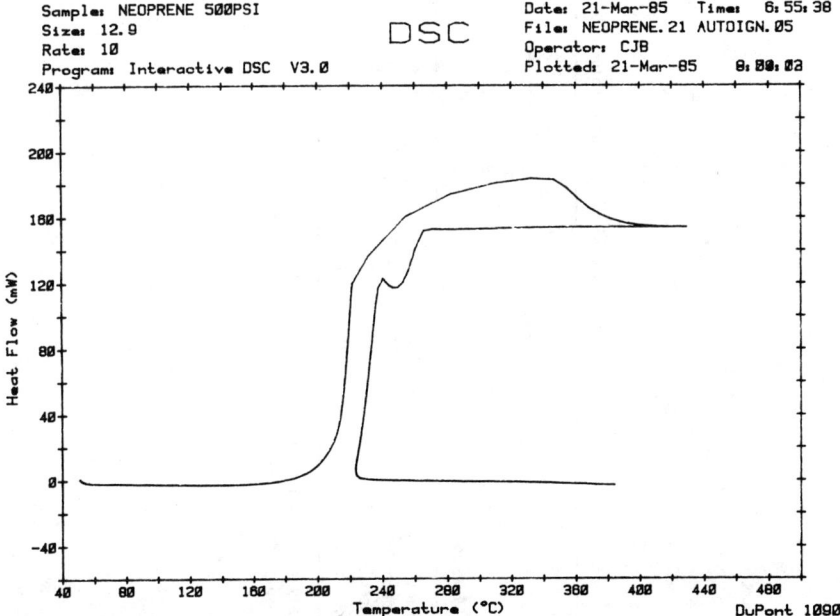

FIG. 5—*Thermogram for 12.9-mg sample of Neoprene® at 3.4 MPa. Note that AIT has now occurred at 209.4°C.*

TABLE 2—Comparative values for AIT's determined by ASTM G 72 and PDSC.

Material	AIT, K											
	G 72 Pressure, MPa						PDSC Pressure, MPa					
	0.68	SD[a]	3.4	SD	6.8	SD	0.68	SD	3.4	SD	6.8	SD
Nylon 6/6	584	b	520	8.8	475	0.4	618	1.1	613	5.2	543	3.0
Neoprene®	472	b	464	3.9	457	0.4	571	3.8	476	1.7	468	0.6
Fluorel E-2160®	>698	b	599	5.9	596	3.5	735	8.7	633	1.7	622	1.0
Vespel SP-21®	>698	b	625	4.7	610	2.0	759	5.0	694	7.1	671	6.5

[a] SD = standard deviation.
[b] Single data point.

the same in both methods. At all pressures, Vespel SP-21® had the highest AIT and was followed by Fluorel E-2160®, nylon 6/6, and Neoprene® in decreasing order.

The most logical explanation for the higher AIT's in the PDSC method probably lies in the difference in the sample cell configuration. The sample is essentially totally encapsulated in glass in the ASTM cell whereas the sample in the PDSC cell is entirely surrounded by various metals that may interact with the gaseous products from the test sample. This interaction would reduce the effective concentration of the decomposition and outgassing products, thus requiring a higher temperature to produce the necessary flammable mixture. In support of this argument, it is interesting to note that the two halogenated materials in this study, Neoprene® and Fluorel E-2160®, exhibited the least difference in AIT's between the two methods. The PDSC sample chamber is constructed of constantan and silver, and the halogenated decomposition products were observed to have reacted with the metal surfaces. This probably reduced the surface reactivity to the other decomposition products. In fact, these halogenated materials were found to effectively reduce the uselife of the test cell considerably.

Using the PDSC method, it was found that other information became

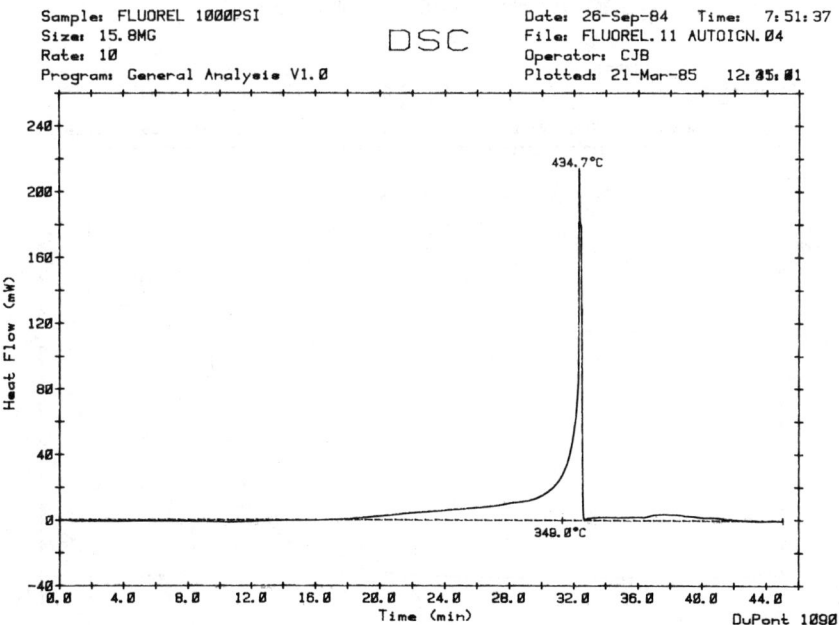

FIG. 6—*Thermogram for a 15.8-mg Fluorel E-2160® sample at 6.8 MPa. The AIT was 349.0°C and the peak temperature was 434.7°C. Note the oxidative stability at the lower temperatures.*

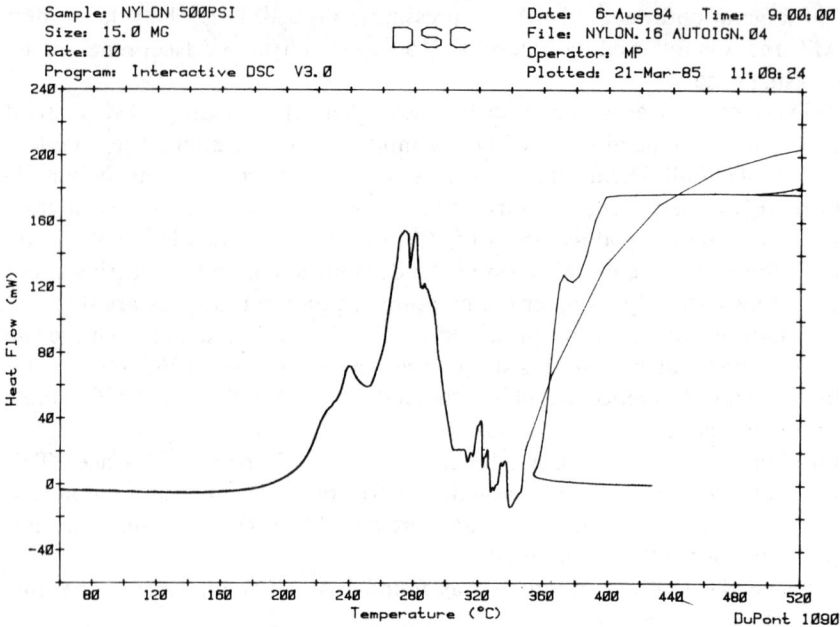

FIG. 7—*Thermogram for 15.0-mg sample of Nylon 6/6 at 3.4 MPa. Note the extensive oxida-tion reactions that occurred in the 200 to 340°C range. AIT was 346.1°C.*

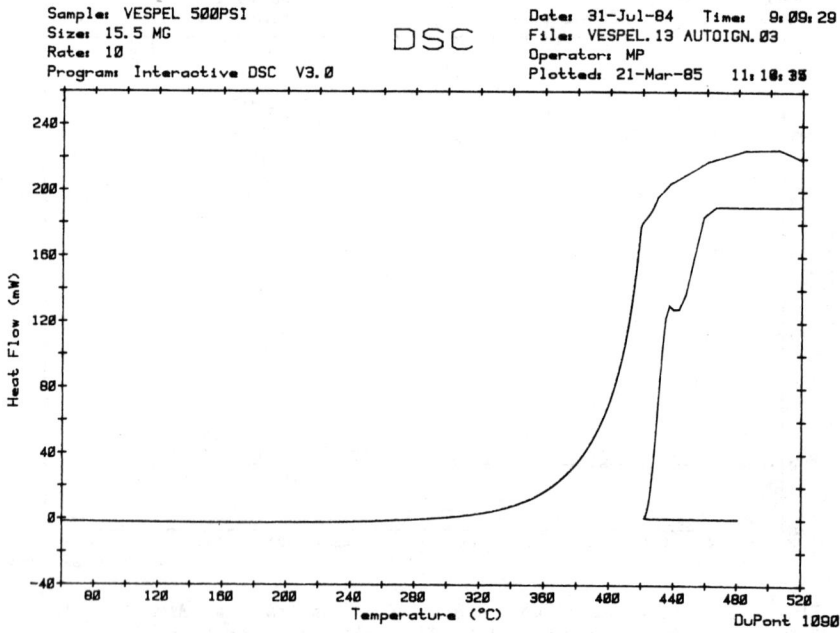

FIG. 8—*Thermogram for 15.5-mg sample of Vespel SP-21® at 3.4 MPa. Note the oxidative stability at the lower temperatures. AIT was 426.5°C.*

available that may be of interest in the selection of materials for oxygen service. The first, and most obvious, is the ability to determine the onset of oxidation of the material. This is important in the determination of the maximum use temperature of a material. The second data point available is the heat of combustion that may be calculated by several of the analysis programs written for the instrument. Typical thermograms for Fluorel E-2160®, nylon 6/6, and Vespel SP-21® are shown in Figs. 6, 7, and 8, respectively.

Conclusions

The comparative determination of the AIT of Neoprene®, Vespel SP-21®, Fluorel E-2160®, and nylon 6/6 by ASTM G 72 and a PDSC method revealed that the ASTM method gave lower AIT's for all four materials. However, the ranking of the materials was identical in both methods. Additional information, onset of oxidation, and heat of combustion, may be obtained using the PDSC data recording and analysis capabilities.

Acknowledgments

This work was performed under the sponsorship of the National Aeronautics and Space Administration (NASA) and the BOC Group, Inc. Neither the U.S. government, the BOC Group, Inc., or any person acting on their behalf assumes any liability resulting from the use of the information contained in this paper, or warrants that such use will be free from privately owned rights. The use of company trademarks or tradenames does not constitute an endorsement of these materials, but is used only to simplify their identification.

Reference

[1] Nihart, G. J. and Smith, C. P., "Compatibility of Materials with 7500 psi Oxygen," DDC AD 608260, AMRL-TDR-64-76, Union Carbide Corp., Tonawanda, NY, Oct. 1964.

Jun'ichi Sato[1] and Toshisuke Hirano[2]

Behavior of Fire Spreading Along High-Temperature Mild Steel and Aluminum Cylinders in Oxygen

REFERENCE: Sato, J. and Hirano, T., **"Behavior of Fire Spreading Along High-Temperature Mild Steel and Aluminum Cylinders in Oxygen,"** *Flammability and Sensitivity of Materials in Oxygen-Enriched Atmospheres: Second Volume, ASTM STP 910*, M. A. Benning, Ed., American Society for Testing and Materials, Philadelphia, 1986, pp. 118–134.

ABSTRACT: Effects of the metal temperature and the metal surface condition on the fire spread rate along mild steel and aluminum cylinders have been studied experimentally for wide variations of the oxygen pressure, up to 10 MPa, and the test piece temperature, up to the melting points. Based on the experimental results, the mechanisms by which fire spread is controlled, and the dependence of the fire spread rate on the metal temperature and the metal surface condition have been discussed.

The experimental results show that the fire spread rate does not always increase with the increase of the metal temperature, but the two cases, the increase or the decrease, exist corresponding to the ranges of the oxygen pressure and the metal temperature. The curious dependency of the spread rate on the metal temperature is attributable to the variation of the heat transfer rate from the molten mass to the unburned solid caused by the change of the contour of the boundary surface between the molten mass and the unburned solid metal. The effect of the surface condition on the fire spread rate is small. In this case, the variation of the fire spread rate is mainly attributable not to the variation of the surface reaction rate or the heat transfer rate at the solid metal just above the molten mass, but to the variation of the heat transfer rate from the molten mass to the unburned solid.

KEY WORDS: metals, steels, aluminum, oxygen, combustion, fire accident, oxygen fire, fire spread, burning behavior, combustion temperature, flammability

In a high-pressure oxygen atmosphere, most metals are flammable. Therefore, serious accidents of various chemical plants and rocket motors have

[1]Senior researcher, Research Institute, Ishikawajima-Harima Heavy Industries Co., Ltd., Toyosu, Koto-ku, Tokyo 135, Japan.
[2]Professor, Department of Reaction Chemistry, The University of Tokyo, Hongo, Bunkyo-ku, Tokyo 113, Japan.

been caused by burning metal parts of oxygen systems [1,2]. At these metal fire accidents, ignition occurs because of the heating of metal parts by the friction between moving and stationary parts or other heat sources such as hot burnt products. After ignition, fire spreads along the metal parts of oxygen systems. In these cases, temperature of the metal parts around the ignition point is high, and their surfaces are usually covered with metal oxides. Thus, a detailed study on the fire spreading along the high-temperature metal pieces covered with the oxides is indispensable in order to prevent the metal fire accidents or to minimize the damage caused by them.

Several studies have been carried out on the fire spreading along metal pieces [3-17]. In most of these studies, the effects of the oxygen pressure or test piece dimension on the spread rate were examined, and several facts, which would be useful to understand the general characteristics of metal fire, have been found. Two typical aspects of the fire spread exist; one is the case of iron, and the other is the case of aluminum. For the case of the iron, the exothermic chemical reaction between iron and oxygen occurs in the molten iron oxide mass formed on the burning end. The fire spread rate is defined as the metal melting rate into the molten mass. For the case of the aluminum, the exothermic chemical reaction between aluminum vapor and oxygen occurs in the gas phase around the molten aluminum mass formed on the burning end. The fire spread rate is defined as the melting rate of the solid aluminum into the molten mass. Some analyses were performed for the prediction of fire spread based on the experimental results, and useful information for understanding the fire spread mechanisms has been obtained [12,13,15-17].

Although these studies have yielded much useful information on the metal fire spread, there still remains many problems to be solved. As described above, one remaining problem to be solved is the fire spread along the high-temperature metal pieces. For understanding this, the effects of the oxide layer formed on the metal pieces on the fire spread also have to be examined, because the surface of the high temperature metal pieces is usually covered by its oxide. Thus, in the present study, experiments of fire spread along high-temperature mild steel and aluminum cylinders have been carried out by considering the metal surface condition. Based on the experimental results, the mechanisms by which the fire spread is controlled has been discussed.

Experimental Procedure

Figure 1 shows a schematic of the experimental apparatus used in the present experiments. The cylindrical high-pressure oxygen chamber, which was the same as that used in our previous experimental studies, was used for these experiments [13-16]. In the present experiments, a small high-pressure heating chamber of 122 mm in inner diameter and 240 mm in length, which had an electrical heater and could heat up the test pieces to 1600 K, was inserted in the high-pressure oxygen chamber. In order to prevent the test

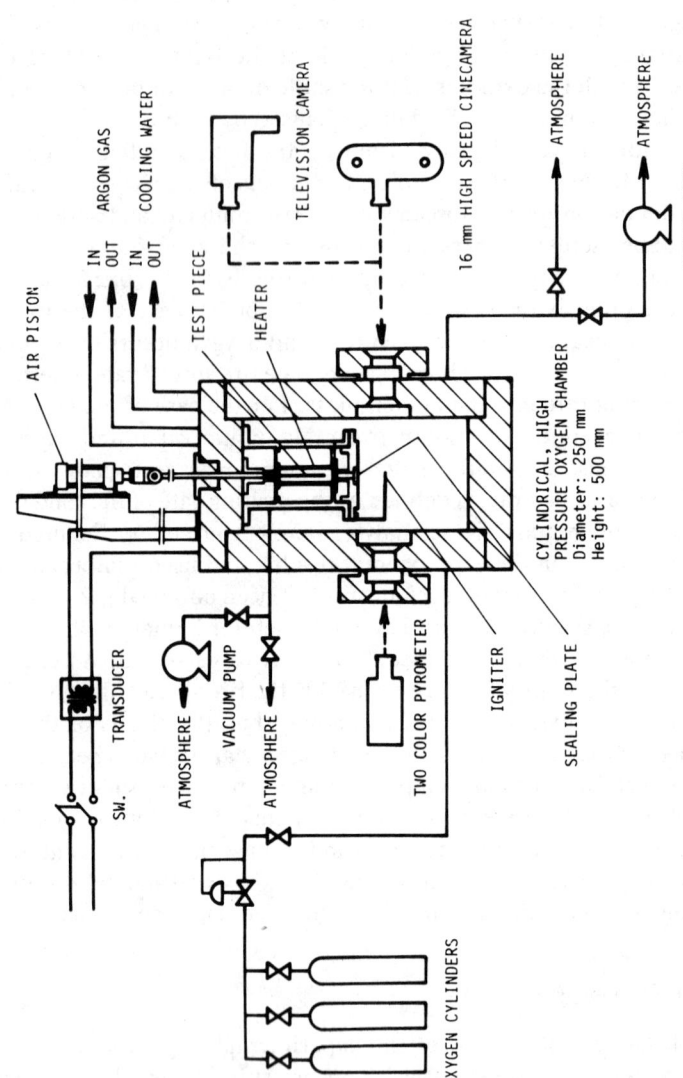

FIG. 1—*Schematic of experimental apparatus.*

pieces from reaction with the oxygen during the heating, the heating chamber was filled up by inert gas. The inert gas used was argon. A small hole, which can be closed or opened by an automatically movable plate, was on the bottom of the heating chamber.

The test pieces, cylinders of metals, were mounted vertically on a supporting system, which was moved by an air piston. At the heating time, the test piece was heated up in the heating chamber. The temperature of the heating chamber was measured by three $Pt/87\% Pt$-13% for rhodium (Rh) thermocouples. When the test piece was heated up to a predetermined value, the small hole was opened, and then the test piece was ejected from the heating chamber into the high-pressure oxygen atmosphere by the air piston. The purity of the oxygen atmosphere was about 99.6% O_2 (others: nitrogen and argon), and its dew point was under 210 K.

Ignition of the test piece was performed by an igniter composed of a matchhead and magnesium ribbons. The igniter was located at the end of the test piece stroke and burned by contact with the end of the hot test piece. As soon as the end of the hot test piece was contacted with the igniter, the test piece was ignited at its bottom end. By using this ignition system, the test piece could be ignited without the decrease of the test piece temperature, and the effect of the ignition procedure on the fire spread phenomena could be confined within a few centimetres from the ignition end.

The test pieces used were cylinders of mild steel and aluminum. The minor components of the mild steel were 0.069% carbon, 0.015% phosphorus, 0.013% silicon, 0.37% manganese, and 0.014% sulfur, and those of the aluminum were 0.75% silicon + iron, 0.1% zinc, 0.05% manganese, and 0.1% copper. By comparing the observed spread rates with those along pure iron and aluminum cylinders, the influence of these minor species in the test pieces on the spread phenomena was confirmed to be negligible. The test piece diameter was 3 mm and the length was 120 mm. Several kinds of surface condition of the test piece were examined. For the mild steel, surfaces covered with thin iron oxide, thick iron oxide, and gold plating were tested. For the aluminum, surfaces covered with thin aluminum oxide and thick aluminum oxide by means of anodizing. The thickness of the gold plating and the anodized plating were about 3×10^{-3} and about 8×10^{-3} mm, respectively.

Upward fire spread, which is stable and preferable for discussing the flame spread mechanisms than downward fire spread, has been examined. The fire spread phenomena were recorded by using a high speed 16-mm cine-camera and a television with a video recorder. The temperature of the burning region of each test piece was measured by a two color pyrometer and recorder.

Experimental Results

After the ignition of the test piece at its bottom end, the burning region began to spread toward the top end. The appearance of the fire spread along high-temperature test pieces, even for the test pieces whose temperature is close to the melting point, is basically the same as that for the test pieces at room temperature that was previously reported [13–16]. In order to discuss the overall characteristics of fire spread phenomena, mean (time averaged) fire spread rate $\bar{V}$ was obtained from the position-time diagram of the leading edge of the molten mass formed on the bottom end of the test piece.

Mild Steel

When the burning region spreads upward along a steel cylinder, a molten iron oxide mass, which is the luminous spherical part at the bottom end of the test piece, is formed. As the burning region spreads, the molten mass volume increases. When the molten mass becomes too large to be supported, it drops. Just after the most part of the molten mass drops, the remaining molten mass volume starts increasing again. This behavior repeats at regular intervals [13–16].

The variations of the mean fire spread rate $\bar{V}$ with the test piece temperature T_u along the mild steel cylinder are shown in Fig. 2. For the oxygen pressures of $p_o = 1$ and 2 MPa. $\bar{V}$ decreases with the increase of T_u and reaches a minimum at $T_u = $ about 750 K, beyond which $\bar{V}$ increases with T_u. For high pressure of oxygen, $p_o = 5$ and 10 MPa, $\bar{V}$ increases with T_u. When the test pieces of 1573 K were ejected into the oxygen atmosphere without the igniter, auto-ignition of the test pieces occurred. Therefore, $T_u = 1573$ K is considered to be an auto-ignition temperature of the mild steel.

In order to understand the effects of the surface condition on the fire spread rate, fire spread experiments for the test pieces covered with gold plating and that covered with oxide layer have been performed at room temperature. The oxide layer was formed by two different procedures. One procedure is that the oxide layer was formed by ejecting the test piece at $T_u = 773$ K from argon into high-pressure oxygen and cooling down to the room temperature. The thickness of the oxide layer was measured by a microscopy. The oxide layer thickness was 2 to 4×10^{-3} mm, and its color is black. The other procedure is that the oxide layer was removed by the sandpaper just before the experiments, and then the oxide layer was formed at room temperature in short time. The color of the surface is the same as that of the virgin metal surface. The thickness of the oxide layer formed in this procedure (estimated: 2.5×10^{-5} mm) is much thinner than that formed in the former procedure. The effects of the surface condition on the fire spread rate are shown in Fig. 3. The spread rate for the cylinders with the gold plating is higher than that for others. The spread rate for the cylinders with the thick oxide layer is the same as that for cylinders with the thin oxide layer.

Figure 4 shows the dependence of the molten mass temperature T_b, the temperature of the burning region, on the oxygen pressure p_o. The temperature of the molten mass T_b increases slightly with the increase of the oxygen

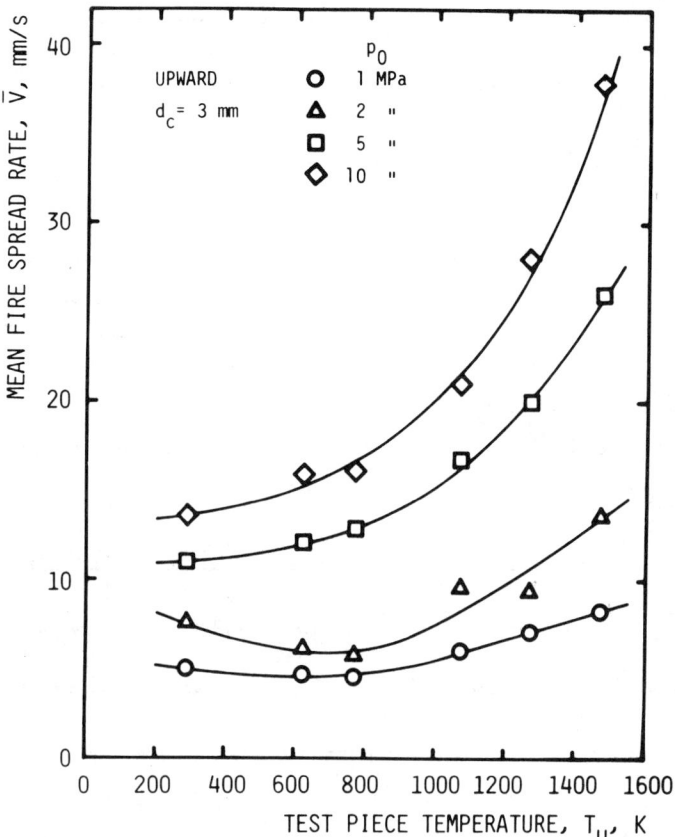

FIG. 2—*Variations of the fire spread rate with the test piece temperature along the mild steel cylinder.*

pressure p_o. T_u and the surface condition have no effect on T_b. The value of the molten mass temperature obtained in this experiment is consistent with that obtained in our previous studies by different means [*13-16*].

Aluminum

When the burning region spreads upward along the aluminum cylinder, a molten aluminum mass, which is surrounded by a flame, is formed. As the burning region spreads upward, the volume of the molten aluminum mass increases. During this period, the surface of the molten mass is seen to be smooth. When the molten aluminum mass becomes large, the molten mass starts boiling and fragments of molten aluminum scatter into the surroundings. At the same time, the smoke density near the molten mass and the light

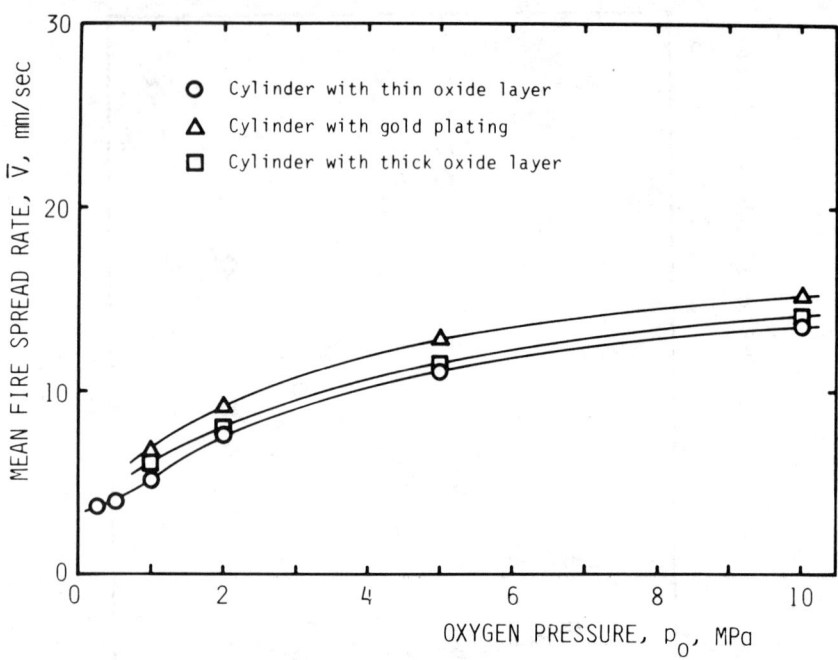

FIG. 3—*Effects of the surface condition on the fire spread rate for the mild steel cylinder.*

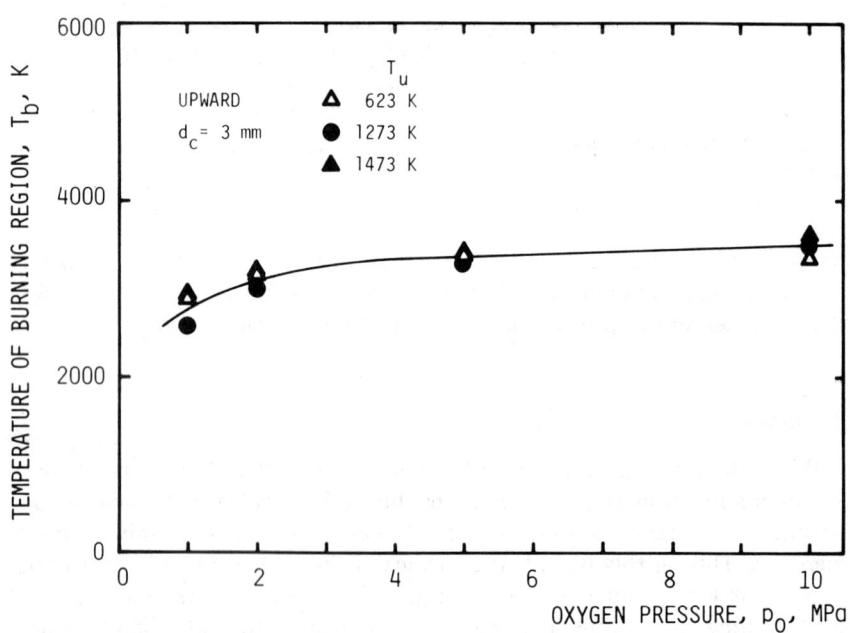

FIG. 4—*Dependence of the molten mass temperature on the oxygen pressure for the mild steel cylinder.*

emission caused by combustion are seen to increase. When the molten mass becomes too large to be supported, it drops. After the molten mass detaches from the bottom end as a droplet, the remaining molten mass starts increasing again [16].

Figure 5 shows the position-time diagram of the leading edge of the molten mass. In this figure, the circle represents the detachment of the molten mass, and the triangle represents the start of boiling of the molten mass. When T_u is the room temperature, the local fire spread rate dx/dt of the nonboiling period (the time between the circle to triangle) is large compared with that of the boiling period [16]. On the other hand, for $T_u = 873$ K, dx/dt of the nonboiling period is not so different with that of the boiling period. Figure 6 shows the variations of $\bar{V}$ with the oxygen pressure p_o. For $T_u = 293$ K, as p_o is increased, $\bar{V}$ increases to a maximum and decreases. Then $\bar{V}$ becomes a minimum and increases again. On the other hand, for the larger values of T_u, $\bar{V}$ increases with p_o monotonically. Since $\bar{V}$ is the time averaged dx/dt, the curious dependency of $\bar{V}$ on p_o for the room temperature value of T_u is explained by the variation with p_o of the ratio of the nonboiling period, which has large dx/dt, and boiling period, which has small dx/dt [16]. For the larger values of T_u, dx/dt of the nonboiling period and that of boiling period are similar. Therefore, the variation of the ratio of the nonboiling period and boiling period with p_o does not result in the curious dependency of $\bar{V}$ with p_o as occurs when T_u is room temperature.

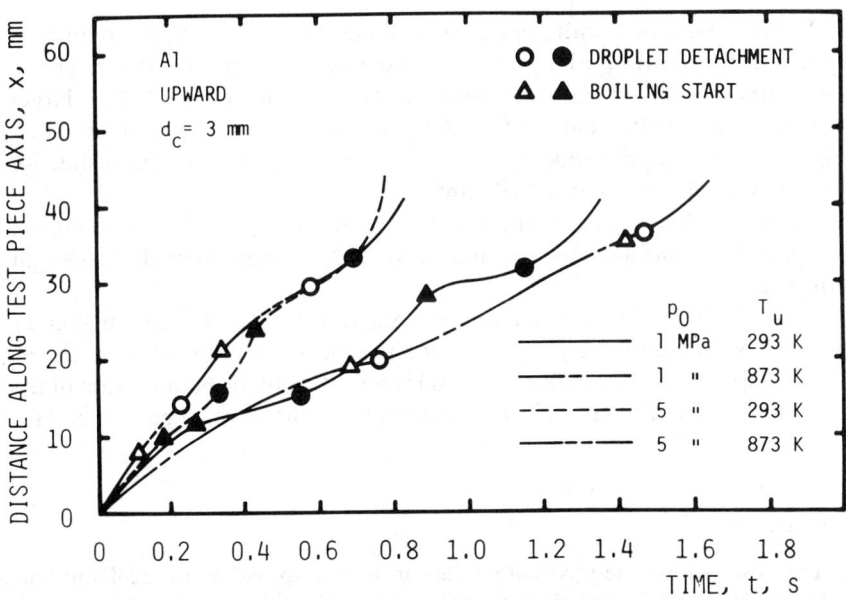

FIG. 5—*Position-time diagram of the leading edge of the molten mass for the aluminum cylinder.*

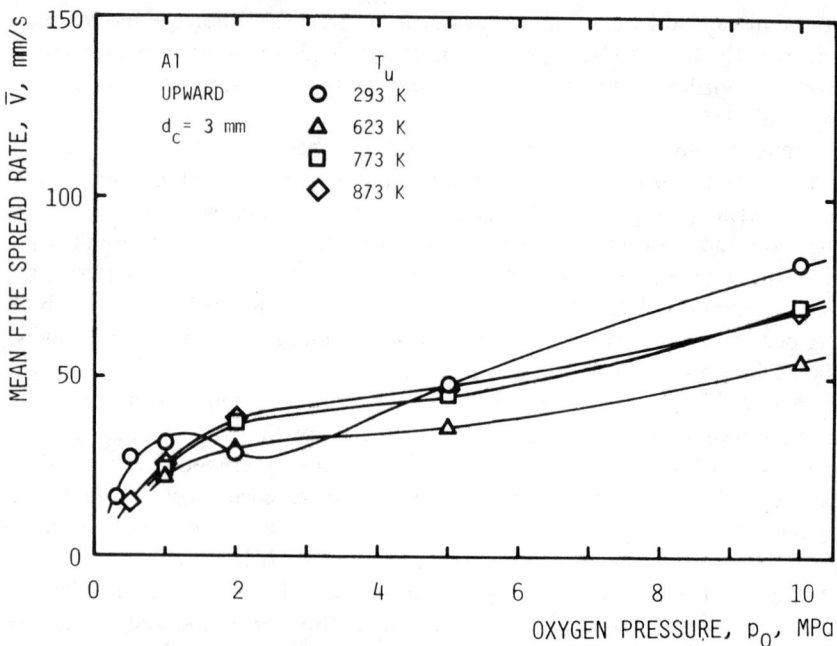

FIG. 6—*Variations of the fire spread rate with the oxygen pressure for the aluminum cylinder.*

The variations of $\bar{V}$ with the test piece temperature T_u along the aluminum cylinders are shown in Fig. 7. As T_u is increased, $\bar{V}$ decreases and reaches a minimum, beyond which $\bar{V}$ increases. At $T_u = 293$ K, $\bar{V}$ for 1 MPa is larger than that for 2 MPa, and the lines for 1 and 2 MPa cross each other. From Fig. 6, the curious dependency of $\bar{V}$ on p_o for $T_u = 293$ K is responsible for the crossing of the 1- and 2-MPa lines.

Figure 8 shows the effects of the surface condition on the fire spread rate $\bar{V}$. The anodized plating, which is composed of the aluminum oxide, has slight effects on $\bar{V}$.

Figure 9 shows the variation of the temperature of the burning region T_b with the oxygen pressure p_o. p_o, T_u, and the surface condition has no effects on T_b. The value of T_b (3300 to 3400 K) is very close to the boiling point of the aluminum oxide (about 3250 K at atmospheric pressure, slightly increases with p_o).

Discussion

Figure 10 shows the physical models of the fire spread along steel and aluminum cylinders. It is obvious that the fire spread rate is proportional to the heat transfer rate to the unburned solid. Based on this fact, the brief analyses

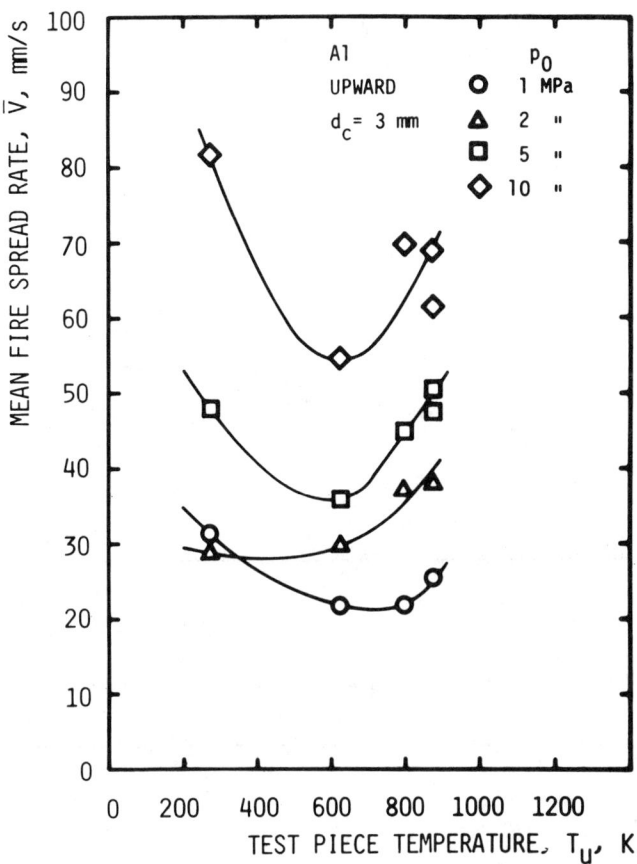

FIG. 7—*Variations of the fire spread rate with the test piece temperature along the aluminum cylinder.*

have been performed in our previous studies [12,15-17]. From the results of these, the mean fire spread rate $\bar{V}$ for the steel cylinder is expressed as

$$
\bar{V} = \frac{1}{2} \left[\frac{\bar{S}}{(\pi d_c^2/4)} \times \frac{B_C}{\rho[c_p(T_m - T_u) + q_L]} \right.
$$

$$
+ \left\{ \left(\frac{\bar{S}}{(\pi d_c^2/4)} \times \frac{B_C}{\rho[c_p(T_m - T_u) + q_L]} \right)^2 \right.
$$

$$
\left. \left. + \frac{16}{d_c} \times \frac{B_A}{\rho[c_p(T_m - T_u) + q_L](T_m - T_u)} \right\}^{1/2} \right]
$$

(1)

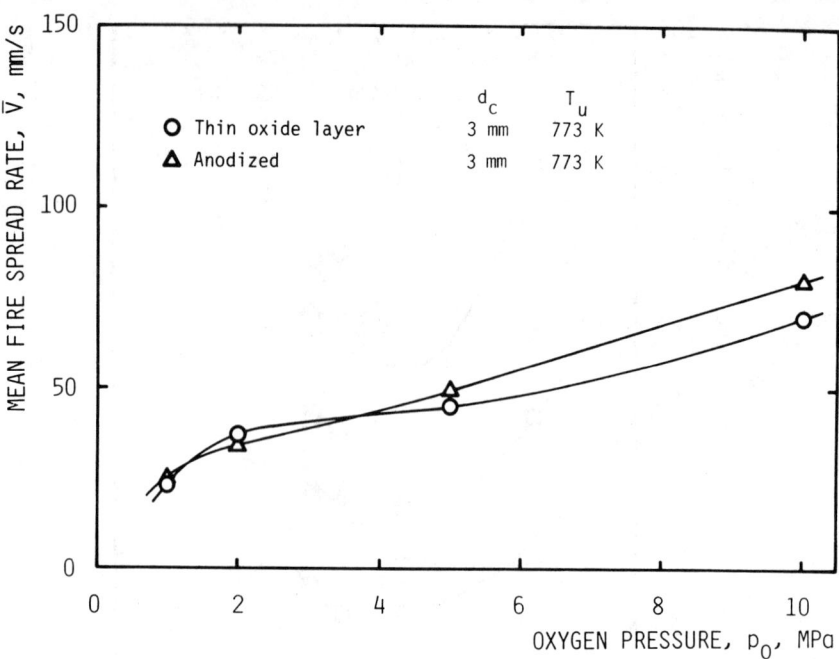

FIG. 8—*Effects of the surface condition on the fire spread rate for the aluminum cylinder.*

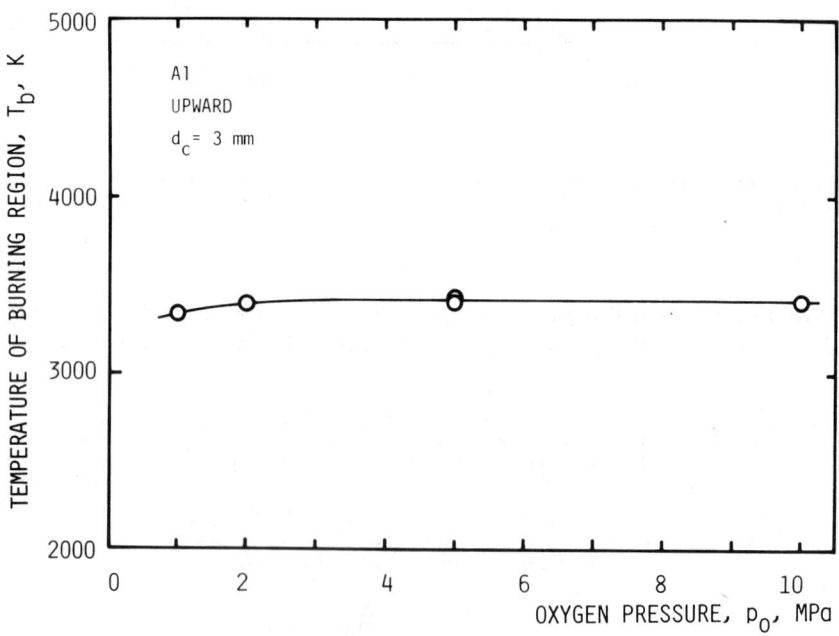

FIG. 9—*Variations of the temperature of the burning region with the oxygen pressure for the aluminum cylinder.*

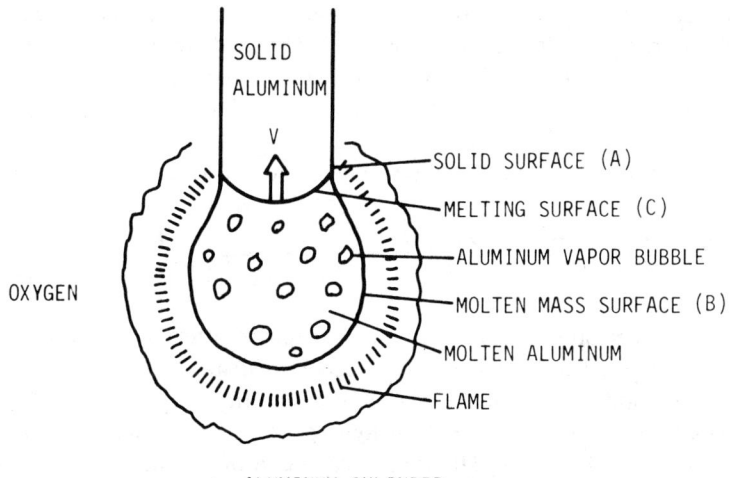

a. ALUMINUM CYLINDER

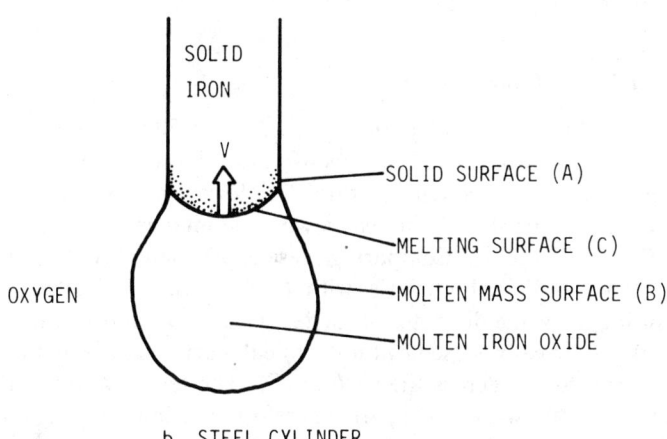

b. STEEL CYLINDER

FIG. 10—*Physical model of the fire spread along aluminum and steel cylinders.*

where $\bar{S}$ is the surface area of the molten mass, d_c the cylinder diameter, T_m the melting temperature, q_L the latent heat, ρ the density, and c_p the heat capacity. In this equation, B_A and B_C are the important parameters in a discussion of the effects of the metal piece temperature and the surface condition on the flame spread rate. B_A is a quantity proportional to the oxygen adsorption rate per unit area on the surface (A) and the ratio of the heat transferred into the solid to the heat released at the surface (A). B_C is a quantity proportional to the heat released by the chemical reaction in the molten mass and the ratio of the heat transferred across the surface (C) to the heat released in the molten mass. The dominant mode of the heat transfer at the surface (C) is convection of the molten mass. Comparing the heat transfer rate across the surface (A) and the surface (C) to the solid steel, the latter is much larger than the former. It implies that in Eq 1 the term involving B_C is much larger than the term involving B_A [15–17].

For the aluminum cylinder, the mean fire spread rate $\bar{V}$ is expressed as

$$\bar{V} = \frac{4A_A}{\delta_c d_c} \times \frac{(T_f - T_a)}{\rho[c_p(T_m - T_u) + q_L]} + \frac{A_C(T_l - T_m)}{\rho[c_p(T_m - T_u) + q_L]} \quad (2)$$

where T_f is the flame temperature, T_a the temperature at the solid surface close to the flame leading edge, δ_c the quenching distance, and T_l the temperature of the molten aluminum mass. A_A is a quantity proportional to the thermal conductivity, and A_C is a quantity depending on the heat transfer efficiency at the surface (C). The dominant mode of the heat transfer at the surface (C) is convection. The heat transfer rate across the surface (C) is much larger than that across the surface (A); the term involving A_C is much larger than that involving A_A [16].

Effects of Metal Temperature

For the mild steel cylinder, in Eq 1, $q_L/c_p = 582°$ and B_A is negligibly small compared to B_C. So, if B_C is constant, $\bar{V}$ increases gradually with the increase of the test piece temperature T_u. However, it is found from Fig. 2 that $\bar{V}$ decreases or slightly increases with the increase of T_u except for the large value of p_o and T_u. Comparing these experimental results with Eq 1, it is found that S or B_C or both varies with T_u. From the movie films, S is almost constant for the same diameter of the test piece if T_u and p_o vary. B_C is proportional to the heat released by the chemical reaction in the molten mass and the heat transferred across the surface (C). The temperature of the molten mass does not vary with T_u (Fig. 4). Therefore, it is inferred that the variation of the heat transfer rate across the surface (C) is responsible for the change of B_C.

The heat transfer across the surface (C) is governed by the area and contour of the surface (C). Figure 11 shows the photographs of the test pieces that were extinguished by argon during burning at $p_o = 2$ MPa and removed the oxide formed on the burning end. It is found that the contour of the surface (C) in Fig. 10 changes with T_u. The area of the surface (C), first, decreases with the increase of T_u and then becomes constant. Since the heat transfer across the surface (C) is proportional to its area, B_C decreases with the increase of T_u up to a certain value of T_u, beyond which B_C becomes constant. Therefore, the variation of $\bar{V}$ with T_u can be explained by considering the variation of $B_C/[c_p(T_m - T_u) + q_L]$ with T_u.

For the aluminum cylinder, the term involving A_A in Eq 2 is negligibly

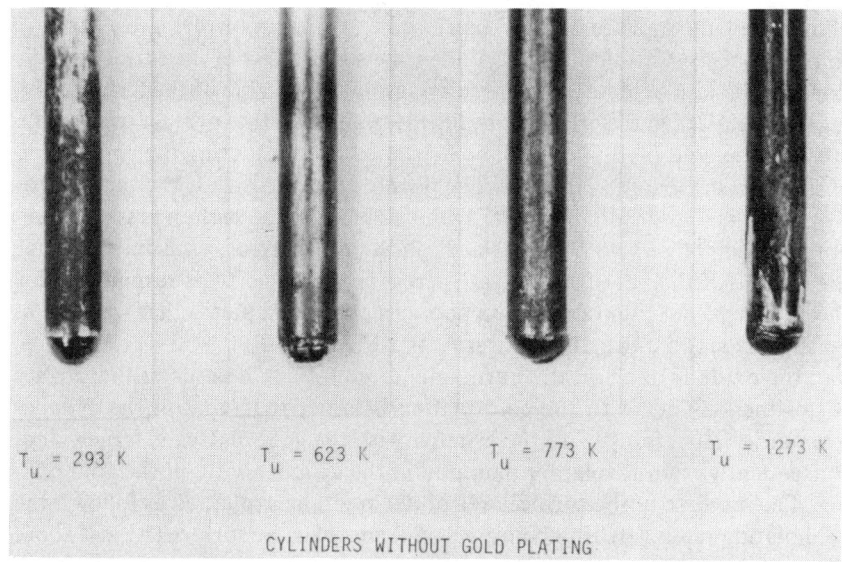

FIG. 11—*Contour of the boundary surface between the molten mass and the unburned solid metal for the mild steel cylinder*; $p_o = 2$ MPa.

small compared to that involving A_C, and $q_L/c_p = 442°$. If A_C is constant, $\bar{V}$ increases with the increase of the test piece temperature T_u. However, from the experiments, $\bar{V}$ decreases and then increases with the increase of T_u (Fig. 7). Comparing the experimental results with Eq 2, it is inferred that this phenomena is attributable to the variation of A_C. A_C varies with the ratio of the nonboiling period to the boiling period and with the change of the surface (C) in Fig. 10. In Fig. 5, the local spread rate dx/dt of the nonboiling period for T_u = room temperature is larger than that for T_u = 873 K. From the discussion for the case of the mild steel, it is inferred that this is also caused by decrease in the area of the surface (C) with the increase of T_u. The decrement of the area of the surface (C) causes the decrease of A_C. Another cause for variation of A_C is the effect of the ratio of the nonboiling period to the boiling period. For large values of T_u, dx/dt of the nonboiling period is not so different with that of the boiling period. Therefore, A_C does not vary so much with the ratio of the nonboiling period to the boiling period but varies with the change of the surface (C). Thus, the variation of $\bar{V}$ with T_u can be explained by considering the variation of the term involving A_C in Eq 2 caused by the change of the surface (C).

Effects of the Surface Condition

Figures 3 and 8 show the effects of the surface condition on the fire spread rate. In Eqs 1 and 2, the effects of the surface condition for mild steel and aluminum cylinders are directly expressed by the term involving B_A and A_A, respectively. For the aluminum cylinder, the slight effect of the surface condition may be well explained by considering the small value of the term involving A_A in Eq 2. On the other hand, for the mild steel, the increase of $\bar{V}$ by the effect of the gold plating can not be explained by considering the small value of the term involving B_A in Eq 1. The possible mechanisms is the variation of $\bar{S}$ or B_C. It is found from the experiments that $\bar{S}$ and the molten mass temperature T_b do not vary with the surface condition. Therefore, it is inferred that the change of the heat transfer rate across the surface (C) is responsible for the change of B_C. Figure 12 shows photographs of test pieces that were extinguished by argon during burning at $p_o = 2$ MPa and $T_u = 773$ K, which then had the oxide formed on the burning end removed. The area of the surface (C) of the cylinder with the gold plating is larger than that of the cylinder without it. Therefore, B_C of the cylinder with the gold plating is larger, and, consequently, $\bar{V}$ of it is larger than that of the cylinder without the gold plating. The increase of the surface area of (C) might be explained as follows: the gold plating melts (melting point: 1336 K) and wets the surface (A) just above the molten mass, and the molten mass climbs up this wetted region by surface tension.

Conclusions

The effects of the metal temperature and the metal surface condition on the fire spread rate along mild steel and aluminum cylinders have been studied

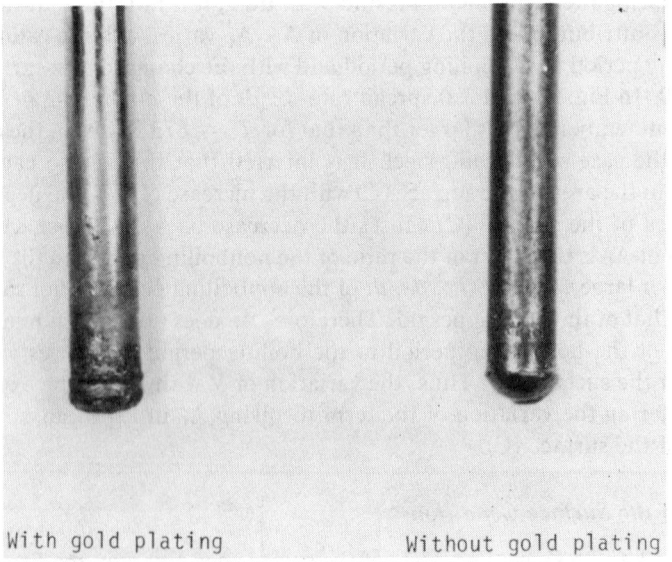

With gold plating   Without gold plating

FIG. 12—*Contour of the boundary surface between the molten mass and the unburned solid metal for the mild steel cylinder;* $T_u = 773$ K, $p_o = 2$ MPa.

experimentally. The experimental results have been discussed based on brief analyses of the fire spread. Some conclusions are summarized as follows:

1. The fire spread rates of the mild steel and aluminum cylinders do not always increase with the increase of the metal temperature, and they may increase or decrease depending on the ranges of the oxygen pressure and the metal temperature.

2. For the low-temperature aluminum cylinder, as the oxygen pressure is increased, the fire spread rate increases to a maximum at p_o = about 1 MPa and decreases. Then the fire spread rate becomes a minimum at p_o = about 2 MPa and increases again with the oxygen pressure. On the other hand, for the high-temperature aluminum cylinder, the fire spread rate increases monotonically with the oxygen pressure.

3. The fire spread rate of the mild steel cylinder increases monotonically with the oxygen pressure for all tested metal temperature.

4. The curious dependency of the fire spread rate on the metal temperature and oxygen pressure is attributable to the variation of the heat transfer rate from the molten mass to the unburned solid caused by the change of the contour of the boundary surface between the molten mass and the unburned solid metal.

5. The effect of the surface condition on the fire spread rate is small. In this case, the variation of the fire spread rate is mainly attributable not to the variation of the surface reaction rate or the heat transfer rate at the solid metal just above the molten mass, but to the variation of the heat transfer rate from the molten mass to the unburned solid.

Acknowledgments

The authors are grateful to Ishikawajima-Harima Heavy Industries Co., Ltd., Turbine & Compressor Division and Sulzer-Escher Wyss Ltd., Thermal Turbomachinery for financial support of this work. They are also indebted to Messrs. K. Takeda and I. Maezawa for their help in conducting experiments.

References

[1] Clark, A. F. and Hust, G., *AIAA Journal*, Vol. 12, No. 4, April 1974, pp. 441–454.
[2] Cataldo, C. E., "LOX/GOX Related Failures during Space Shattle Main Engine Development," NASA TM-82424, National Aeronautics and Space Administration, Washington, DC, 1981.
[3] Grosse, A. V. and Conway, J. B., *Industrial Engineering Chemistry*, Vol. 50, April 1958, pp. 663–672.
[4] Harrison, P. L., *Seventh Symposium (International) on Combustion*, Butterworths Scientific Publications, London, 1959, pp. 913–918.
[5] Harrison, P. L. and Yoffe, A. D., *Proceedings of the Royal Society*, Vol. A261, 1961, pp. 357–370.
[6] Markstein, G. H., *AIAA Journal*, Vol. 1, 1963, pp. 550–562.
[7] Kirschfeld, L., *Archiv Fuer Eisenhüttenwesen*, Vol. 33, 1962, pp. 617–631.
[8] Kirschfeld, L. *Archiv Fuer Eisenhüttenwesen*, Vol. 36, 1965, pp. 823–826.

[9] Kirschfeld, L., *Metallwissenschaft und Technik*, Vol. 21, 1967, pp. 98-102.

[10] Kirschfeld, L., *Archive Fuer Eisenhüttenwesen*, Vol. 39, 1968, pp. 535-539.

[11] Hone, W. and Williams, A., *Journal of the Institute of Fuel*, Vol. 127, 1977, pp. 127-131.

[12] Hirano, T., Sato, K., Sato, Y., and Sato, J., *Combustion Science and Technology*, Vol. 32, 1983, pp. 137-159.

[13] Hirano, T., Sato, Y., Sato, K., and Sato, J., *Oxidation Communications*, Vol. 6, 1984, pp. 113-124.

[14] Sato, K., Sato, Y., Tsuno, T., Nakamura, Y., Hirano, T., and Sato, J., *Fifteenth International Congress on High Speed Photography/Photonics*, The Society of Photo-Optical Instrumentation Engineers, Washington, DC, 1983, pp. 828-832.

[15] Sato, J., Sato, K., and Hirano, T., *Combustion and Flame*, Vol. 51, July 1983, pp. 279-287.

[16] Sato, K., Hirano, T., and Sato, J., ASME-JSME Thermal Engineering Joint Conference, Vol. 4, The Japan Society of Mechanical Engineers, Tokyo, Japan, 1983, pp. 311-315.

[17] Hirano, T., Sato, K., and Sato, J., Fire Dynamics and Heat Transfer, HTD-Vol. 25, The American Society of Mechanical Engineers, New York, 1983, pp. 83-87.

Frank J. Benz,[1] *Randy C. Shaw,*[2] *and John M. Homa*[2]

Burn Propagation Rates of Metals and Alloys in Gaseous Oxygen

REFERENCE: Benz, F. J., Shaw, R. C., and Homa, J. M., **"Burn Propagation Rates of Metals and Alloys in Gaseous Oxygen,"** *Flammability and Sensitivity of Materials in Oxygen-Enriched Atmospheres: Second Volume, ASTM STP 910,* M. A. Benning, Ed., American Society for Testing and Materials, Philadelphia, 1986, pp. 135–152.

ABSTRACT: The average burn rates of several metals and alloys were determined at oxygen pressures between 3.45 and 68.91 MPa (500 and 10 000 psig) and ambient temperature. Several materials were tested at elevated sample temperatures. The test materials were fabricated into solid cylindrical rods and mounted vertically in the test chamber. A magnesium igniter was positioned at the bottom end of each test specimen to promote upward burn propagations.

Nickel 200 and copper 102 could not be ignited at all oxygen pressures tested whereas Monel 400 appeared to ignite but quickly self-extinguished. The other materials tested burned the entire length of the test sample. Aluminum 6061 exhibited the fastest burn propagation rate. Inconel 718 burned slower than aluminum but faster than the stainless steels (Types 304 and 316).

Increasing oxygen pressure generally increased the burn propagation rate of the materials. Increasing the ambient temperature of the test specimens for several materials to approximately 850 K (1070°F) had little effect upon the ignition or burn properties of Nickel 200 or Monel 400. Type 316 stainless steel exhibited an increase in its burn propagation rate at this higher temperature.

KEY WORDS: ignition, burn propagation rate, stainless steels, copper alloys, nickel alloys, aluminum, oxygen compatibility

Nomenclature

C Constant

E Activation energy, kJ

ΔHc Heats of combustion, kJ/g-metal

$\Delta \overset{\circ}{H}c$ Volumetric heat of combustion, kJ/cm^3

[1]Products director, NASA, Johnson Space Center, White Sands Test Facility, Laboratories Test Office, Las Cruces, NM 88004.

[2]Test project engineers, Lockheed/EMSCO, Johnson Space Center, White Sands Test Facility, Material and Component Test Department, Las Cruces, NM 88004.

$\Delta \dot{H}$ Heat release rate, kJ/s

Pox Oxygen pressure, MPa

P Contact pressure, N/m^2

R Gas constant

Rox Oxygen consumption rate per unit area

T Temperature, K

Ts Surface temperature, K

$\overset{\circ}{V}$ Instantaneous burn propagation rate, cm/s

$\dot{V}$ Volumetric burn propagation rate, cm^3/s

$\overline{V}$ Average burn propagation rate, cm/s

n Empirically determined exponent

ℓ Empirically determined exponent

ρ Density, g/cm^3

$\overline{V}$ Average surface speed, m/s

Introduction

A common cause of metal fires in oxygen systems is fire spread from more easily ignitable materials such as hydrocarbon contaminants, component soft goods, or metal particles [1–7]. Thus, potential promoter materials and their adjacent metals are important considerations for the designer of oxygen systems. The National Aeronautics and Space Administration (NASA) White Sands Test Facility (WSTF) is presently involved in developing several test methods for evaluating metals and alloys suitable for oxygen service. One of the methods consists of an apparatus in which metals and alloys are exposed to a high energy promoter, and their fire propagation properties are determined. This paper will describe the apparatus and present data that have been obtained to date.

Background

Two major considerations must be addressed when determining the fire hazards of metals or alloys. First it must be determined whether the material will ignite, which requires knowledge of the minimum energy rate and temperature for ignition. Second, it must be determined whether the material will support combustion after the stimulus of the ignition source has terminated. In this case, the fire spread properties of the material are required. The test method described in this paper evaluates the latter by measuring burn propagation rates of materials.

Previous Studies on Propagation Rates of Metals and Alloys

Kirschfeld [8–16] conducted many experiments in which he measured the average propagation rates $\overline{V}$ of metals and alloys (solid cylindrical specimens). The effects of specimen diameter d and oxygen pressure Pox on $\bar{V}$ were determined. He reported that the effect of d on $\bar{V}$ was generally described by $\bar{V} \propto 1/d$. The effect of Pox on $\bar{V}$ was somewhat more complicated. Metals that burn as a vapor, such as aluminum, exhibited several increases and decreases in $\bar{V}$ as Pox was increased (see Fig. 1). The effect of Pox on $\bar{V}$ for alloys of iron and chromium was related by $\bar{V} \propto Pox^{0.5}$ (see Fig. 2). However multiple slopes were observed for the alloyed steels. Metals, such as magnesium and aluminum, exhibited greater $\bar{V}$'s than iron or the carbon steels. The presence of 2 to 3% chromium and 0.4 to 0.5% molybdenum in low alloy steel appeared to decrease $\bar{V}$ as compared to iron or carbon steel. Steels containing approximately 20% chromium and 10% nickel exhibited greater $\bar{V}$'s than the low alloy steels. Copper, brass, and bronze were ignited but did not propagate the length of the specimen at pressures up to 20.3 MPa (2490 psia) whereas pure nickel could not be ignited.

Harrison and Yoffe [17] conducted similar type propagation tests at pressures below 0.1 MPa (14.7 psia) and also found $\bar{V} \propto Pox^{0.5}$ for iron and iron alloys.

More recently Sato and coworkers [18-23] reported similar results for up-

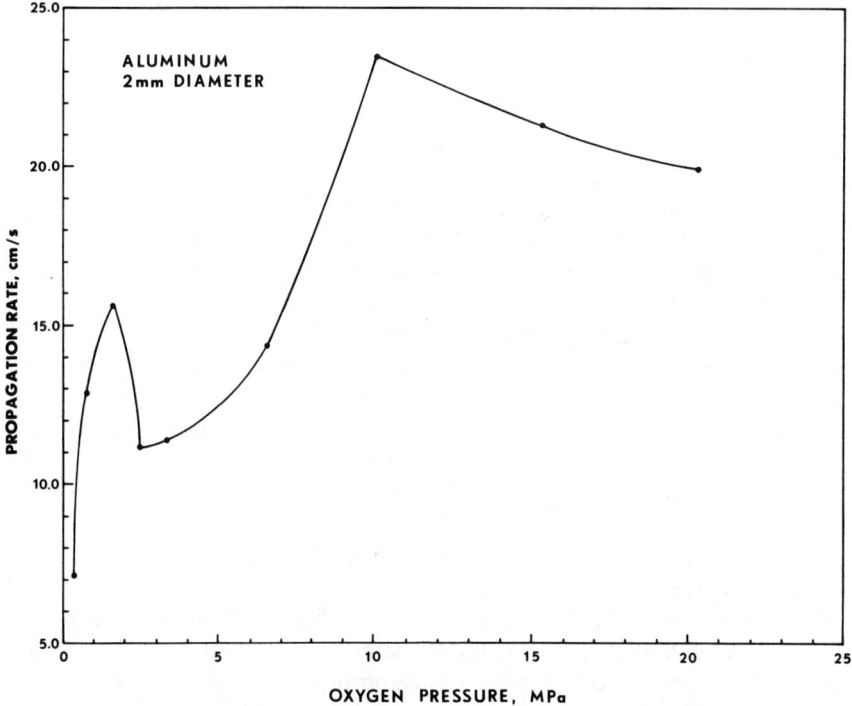

FIG. 1—*Effect of pressure on the burn propagation rate of aluminum* [16].

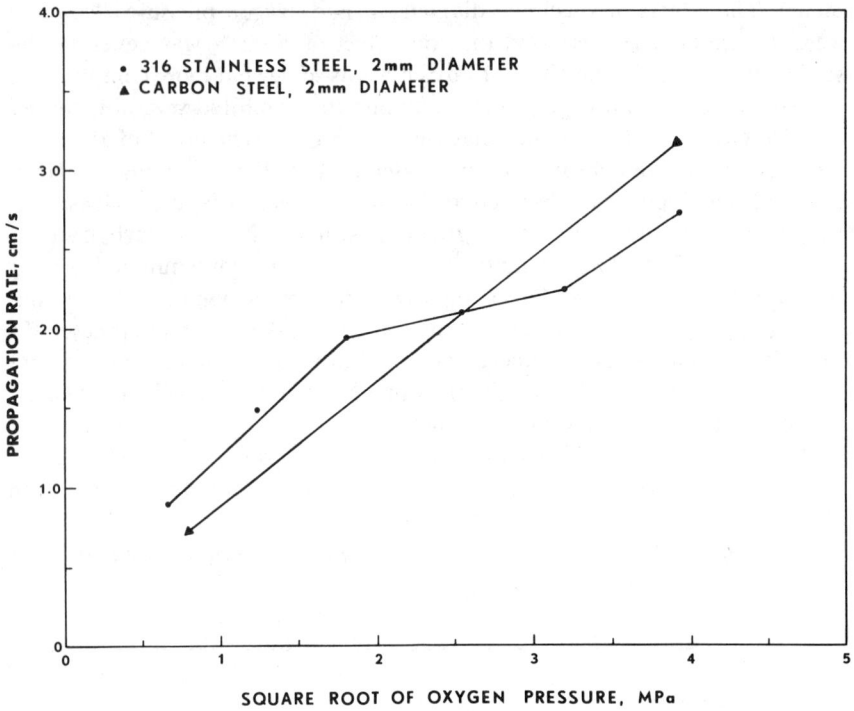

FIG. 2—*Effect of pressure on the burn propagation rates of 316 stainless steel and carbon steel* [9].

ward propagation tests with aluminum and mild steel (solid cylindrical specimens). The effect of Pox on $\bar{V}$ for aluminum showed the same nonuniform relationship as observed by Kirschfeld, and the $\bar{V}$ for mild steel was shown to be proportional to $Pox^{0.4\pm0.05}$.

Sato and coworkers provided high-speed photographic coverage of the burning event. The burning appeared to be oscillatory and corresponded to a continual buildup and detachment of molten material at the bottom of the test specimen. The photographic coverage also showed that aluminum burned in the vapor phase whereas mild steel burned at the surface of the molten mass. The instantaneous burning velocity V in the case of aluminum appeared to increase immediately after detachment of the molten mass and continued to increase until boiling was detected. V then decreased until detachment of the molten mass occurred. In the case of mild steel, V increased slightly just after detachment of the molten mass and then gradually decreased as the size of molten mass increased. V continued to decrease until the next detachment of the molten mass occurred.

Sato and coworkers developed models that described the kinetics and thermal and physical transport properties of the burning events. Several important conclusions about the burning mechanism were drawn when the model was compared to experimental data:

1. The majority of heat from combustion was transferred to the unburned specimen by convection between the molten mass and the solid specimen.

2. The decrease in V during the boiling phase in the case of aluminum was attributed to vapor bubbles that lower the heat transfer between the molten mass and solid specimen.

3. In the case of aluminum, an increase in Pox produced two opposing effects on the heat transfer between the molten mass and solid specimen. There was a greater potential for heat transfer at higher pressures because the boiling temperature of aluminum increased, which, in turn, caused an increase in the temperature of the molten mass. However, at the same time, the increase in pressure decreased the time required to achieve the boiling temperature of aluminum after detachment of the molten mass. This resulted in a decrease in the time available for heat transfer during the phase where the potential for heat transfer was at its maximum. Since V increases as heat transfer increases between the molten mass and the solid specimen, the overall effect of Pox on $\bar{V}$ depended on the pressure effect that was dominating heat transfer to the solid specimen.

4. The effect of Pox on $\bar{V}$ for mild steel was attributed to an increase in the oxidation rate as shown by Eqs 1 and 2 [20]

$$\bar{V} \propto R_{ox} \tag{1}$$

$$R_{ox} = C \, Pox^l T_s^n \exp\,(-E/RTs) \tag{2}$$

Sato et al [22] reported that the effect of Pox on $\bar{V}$ for mild steel was described by an $l = 0.4 \pm 0.5$ or $R_{ox} \propto Pox^{0.4 \pm 0.5}$ for Pox's between 0.3 and 10 MPa (44 and 1450 psig). They concluded from these results that the rate determining step for oxidation was incorporation of oxygen into the oxide layer formed around the molten mass. Hirano et al [20] suggested that the value of l would decrease as Pox is increased. However, they presented no experimental evidence to support their claim.

Data Presented in This Paper

This paper will describe the testing to date that was performed to determine the $\bar{V}$'s of Aluminum 6061, Types 304 and 316 stainless steel, Inconel 718, Monel 400, Nickel 200, and Copper 102 at initial pressures between 3.45 and 68.9 MPa (500 and 10 000 psig) and initial test specimen temperature of approximately 300 K (80°F). A limited number of tests were conducted at an initial sample temperature of approximately 850 K (1070°F). Also presented is a comparison of metals and alloys ranked for oxygen service based on $\bar{V}$ with rankings obtained from frictional heating and particle impact tests.

Test System Description

The test system consisted of a cylindrical stainless steel chamber with an internal volume of approximately 737 cm³ (45 in.³) and capable of being pressurized up to 68.9 MPa (10 000 psig) (see Fig. 3). The chamber was fitted with a copper sleeve and a copper base plate to protect it from burning metal. The test specimen, 0.32 cm (0.125 in.) diameter, 12.7 cm (5 in.) long solid rod, was held at the top by the support assembly and was connected at the bottom to the test specimen heater feedthrough. The promoter consisted of a magnesium sleeve, 0.46 cm (0.25 in.) outside diameter and 0.41 cm (0.16 in.) long, and was held on the test specimen with a press fit. Ignition of the magnesium promoter was accomplished by electrically heating a nichrome wire, which was wrapped around the promoter. For tests at elevated specimen temperatures, the test specimen was electrically heated by providing power to the sample heater feedthrough.

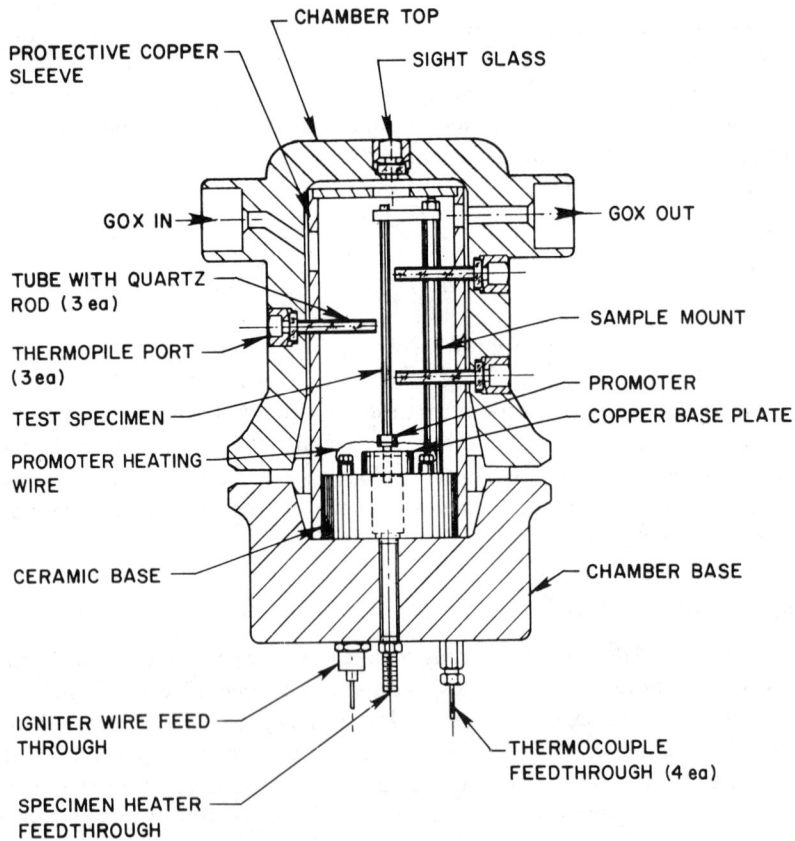

FIG. 3—*Test apparatus.*

Oxygen was supplied to the test chamber via a pumping system from an oxygen accumulator (1.7 m^3). The oxygen purity was checked weekly, and analysis of the oxygen indicated that the purity was between 99.75 and 99.9% by weight. The accumulator provided oxygen at a pressure of 34.5 MPa (5000 psig). When oxygen pressure above 34.5 MPa was required, an oxygen intensifier was employed that delivered oxygen at 68.9 MPa (10 000 psig). All tests were conducted in a static oxygen environment.

The burn propagation rate was determined from the response of three thermopiles (radiant heat detectors) attached to sight glass ports in the wall of the chamber. The sight glass ports were spaced with a vertical distance of 2.54 cm (1 in.) between them and were positioned to provide a direct, horizontal line of sight to the test sample. Each sight glass port was fitted with a 0.48-cm (0.19-in.) inside diameter (ID) tube containing a quartz rod that was positioned inside the chamber between the sight glass and the test sample. Average propagation rates were calculated by comparing the time between the peak responses of the thermopiles and the distance between each thermopile. Propagation rates determined for each test consisted of an average of two values obtained between the three thermopiles.

Other instrumentation consisted of thermocouples for measuring the temperature of the oxygen and test specimen (three locations), a two-color pyrometer (installed at the top of chamber) for measuring the flame temperature, and a bonded strain gage transducer for measuring oxygen pressure.

The data were digitally processed by a microprocessor and stored on a floppy disk. Data from each instrumentation channel were stored every 100 ms. A digital oscilloscope was used when data faster than 100 ms were required.

Results

Selection of a Standard Promoter

A series of tests was conducted with Type 316 stainless steel to determine a suitable promoter that would provide a sufficient and repeatable energy flux density to ignite the test specimen. Only metals were considered as opposed to organic based materials to minimize contamination of the oxygen atmosphere by the combustion products of the promoter. Magnesium and aluminum were chosen as candidate materials. Promoters of the same configuration and dimensions were fabricated from both metals. The differences in the energy release of the two promoters (standard configuration) are given in Table 1, along with the adiabatic flame temperatures. Note that the aluminum promoter provides nearly twice the energy to the test specimen at a higher flame temperature. In Table 2 the $\bar{V}$'s for Type 316 stainless steel are given at two pressures using the aluminum and the magnesium promoters. The results indicated that even though the aluminum promoter was more energetic, the

TABLE 1—*Energy release and flame temperature for the magnesium and aluminum promoters (identical configuration and dimensions).*

Promoter Material	Mass, g	Energy[a] Release, kJ	Adiabatic Flame[b] Temperature, K
Magnesium	0.15	3.7	3623
Aluminum	0.23	7.1	4073

[a]Based on the mass of promoter and assuming complete combustion to magnesium oxide (MgO) or alumina (Al_2O_3).
[b]From Grosse and Conway [25].

TABLE 2—*Propagation rates of Type 316 stainless steel using the magnesium and aluminum promoters.*

Promoter Material	Pressure		Propagation Rates, cm/s (SD)[a]
	MPa	psig	
Magnesium	6.89	1000	1.12 (0.02)
Magnesium	34.46	5000	1.24 (0.05)
Aluminum	6.89	1000	1.07 (0.05)
Aluminum	34.46	5000	1.24 (0.07)

[a]Standard deviation.

effects on the propagation rate of Type 316 stainless steel were insignificant. Magnesium was selected as the material for the standard promoter because it was easier to ignite.

Oxygen Consumption During the Burning Event

Since testing was conducted with a fixed amount of oxygen in the chamber, a drop in oxygen pressure during the burning event could effect the measurement of $\bar{V}$. Several experiments were conducted at oxygen pressures of 6.89 MPa (1000 psig) and above to determine the amount of oxygen consumed during the burning event. The drop in oxygen pressure after burning was completed was less than 4% at 6.89 MPa and must be less than 4% at higher initial oxygen pressures. These results were corroborated by calculating the theoretical consumption of oxygen required to totally burn the test specimen and promoter. The calculated oxygen pressure drop at an initial oxygen pressure of 6.89 MPa was determined to be less than 6%.

Ranking Material Based on Propagation Rate

Burn propagation rates provide a measure of the tendency of a material to self-propagate a fire. However, the relationship between $\bar{V}$ and the susceptibility of a material to ignite should be explored. Testing has indicated that $\bar{V}$ does not characterize the susceptibility of a material to ignite. Table 3 provides a comparison of ranking materials based on $\bar{V}$ with rankings based on ignition by frictional heating [24], and ignition by particle impact [7] in order of decreasing susceptibility to ignition. The results indicate that ranking materials by $\bar{V}$'s do not follow in all cases the ranking of materials by the other tests. This is most evident by the ranking position of copper and Inconel 718 in Table 3. Also, in this study, it was observed that Inconel 718 was harder to ignite than Type 316 stainless steel, even though Inconel 718 exhibited larger $\bar{V}$'s than 316 stainless steel. For example, out of eight tests conducted with Inconel 718 at 6.89 MPa (1000 psig), only four tests resulted in consumption of the entire specimen. In the other four tests, either Inconel 718 did not ignite or burning was limited to less than 50% of the specimen. In the case of Type 316 stainless steel, all tests conducted at the same pressure as Inconel 718 resulted in ignition and total consumption of the test specimens.

However, $\bar{V}$ is an important parameter to consider during fire hazard analysis. $\bar{V}$ describes the combustion event after materials have ignited and, when used with ignition data, provides an overall picture of the hazard involved in selecting a particular material for oxygen service.

A typical example in which $\bar{V}$ data would be valuable is in the selection of materials for an oxygen system where frictional heating can occur. Testing at WSTF (see Table 3) has indicated that most metals and alloys can be ignited by friction heating in oxygen [7]. Even though ignitions cannot be eliminated completely, the severity of a possible fire can be minimized by considering $\bar{V}$'s and heats of combustion (ΔHc) for materials, that is, heat release rates ($\Delta \dot{H}c$). A material that exhibits a large $\Delta \dot{H}c$ will theoretically have a greater potential to heat larger portions of a system than a material that exhibits a small $\Delta \dot{H}c$ for a given system heat loss rate. Table 4 gives $\Delta \dot{H}c$ calculated for three alloys based on $\bar{V}$'s determined in this study (27.57-MPa oxygen pressure) and calculated ΔHc's. Of the three alloys considered, Inconel 718 would be the best choice because it exhibits the smallest $\Delta \dot{H}c$ even though $\bar{V}$ for Inconel 718 was greater than the $\bar{V}$ observed for Type 316 stainless steel. In addition, Inconel 718 requires a greater frictional energy flux for ignition than the other two alloys (Table 3).

Effects of Pressure on Burn Propagation Rates

The effects of Pox on $\bar{V}$ were investigated at Pox's between 3.45 and 68.91 MPa (500 and 10 000 psig) for several metals and alloys (Table 5). Copper 102 and Nickel 200 did not ignite. Monel 400 exhibited partial burning at the three test pressures and less than 30% of the test specimen was consumed; no $\bar{V}$'s could be determined. In the case of Types 316 and 304 stainless steel, partial burns were observed at 3.45 MPa (500 psig) in which less than 40% of the specimens were consumed.

TABLE 3—*Ranking of ignition susceptibility of metals and alloys by various methods.*

Propagation Rate (This Work)	Ranking[a] Criterion $\bar{V}$, cm/s	Propagation Rate [9,15,17]	Ranking[b] Criterion $\bar{V}$, cm/s	Ignition by Frictional Heating, [24]	Ranking[c] Criterion PV Product, N/m²·m/s	Ignition by Particle Impact [7]	Ranking[d] Criterion Target Temperature, K
Copper	no ignition	nickel	no ignition	Copper 102[e]	no ignition	copper	no ignition (700)[f]
Nickel 200	no ignition	iron	0.54	Nickel 200	2.31×10^1	Nickel 200	no ignition (713)[f]
Monel 400	partial burn	carbon steel	0.57	Monel 400	1.47×10^8	Monel 400	no ignition (713)[f]
304 SS	1.19	316 SS	0.82	Inconel 718	1.0×10^8	Inconel 718	575
316 SS	1.22	copper	0.95	304 SS	0.7×10^8	316 SS	475
Inconel 718	1.30	aluminum	8.1	316 SS	0.6×10^8	304 SS	425
Aluminum 6061	11.5	...	...	Carbon Steel	0.6×10^8	Aluminum 6061	400
...	...	...	...	Aluminum 6061	0.07×10^8	...	...

[a]3.2-mm-diameter rod, 20.67-MPa (3000-psig) oxygen pressure.
[b]1.0 mm diameter were 0.1-MPa (1.0-psig) oxygen pressure.
[c]Surface speed, between 10 and 20 m/s, 6.9-MPa (1000 psig) oxygen pressure.
[d]1600-μm aluminum particle, 34.5-MPa (5000-psig) initial pressure, sonic gas velocity.
[e]Failed mechanically.
[f]Maximum test temperature.

TABLE 4—*Fire spread properties of several alloys.*

Alloy	$\bar{V}^a$, cm/s	$\overset{\circ}{V}{}^b$, cm³/s	$\Delta\overset{\circ}{H}c\,^c$, kJ/cm³	$\Delta\dot{H}c^d$, kJ/s
Inconel 718	1.3	0.104	54.3	5.7
316 SS	1.22	0.098	62.0	6.1
Aluminum 6061	11.5	0.920	83.8	77.1

[a]Results taken from this work; 3.2-mm-diameter rods and upward burn propagation.
[b]Volumetric burn rate; $\overset{\circ}{V} = V \cdot A$.
[c]Volumetric heat release (based on complete combustion to common oxides: Fe_2O_3, NiO, Cr_2O_3, Al_2O_3, MoO_2, Ta_2O_5); $\Delta\overset{\circ}{H}c = \Delta Hc \times \rho$.
[d]Heat release rate: $\Delta\dot{H}c = \overset{\circ}{V} \times \Delta\overset{\circ}{H}c$.

TABLE 5—*Effect of pressure on propagation rates.*

Material	Pressure		Number of Tests	Propagation Rate[a]	
	MPa	psig		cm/s	(SD)[b]
316 stainless steel	3.45	500	4	partial burn	
316 stainless steel	6.89	1 000	5	1.12	(0.03)
316 stainless steel	20.67	3 000	2	1.22	(0.0)
316 stainless steel	27.57	4 000	6	1.24	(0.05)
316 stainless steel	48.24	7 000	5	1.44	(0.06)
316 stainless steel	68.91	10 000	4	1.58	(0.06)
304 stainless steel	3.45	500	10	partial burn	
304 stainless steel	20.67	3 000	10	1.19	(0.05)
304 stainless steel	34.46	5 000	1	1.30	(-)
Inconel 718	6.89	1 000	4	1.12	(0.1)
Inconel 718	27.57	4 000	6	1.33	(0.05)
Inconel 718	48.24	7 000	5	1.50	(0.06)
Inconel 718	68.91	10 000	5	1.68	(0.06)
Monel 400	6.89	1 000	1	partial burn	
Monel 400	34.46	5 000	2	partial burn	
Monel 400	55.13	8 000	3	partial burn	
Aluminum 6061	6.89	1 000	4	6.42	(0.6)
Aluminum 6061	13.78	2 000	2	8.85	(0.6)
Aluminum 6061	27.57	4 000	7	13.86	(0.7)
Aluminum 6061	34.46	5 000	2	14.82	(2.9)
Aluminum 6061	48.24	7 000	2	18.93	(1.8)
Aluminum 6061	68.91	10 000	3	24.51	(0.6)
Copper 102	6.89	1 000	2	no ignition	
Copper 102	34.46	5 000	2	no ignition	
Copper 102	55.13	8 000	2	no ignition	
Nickel 200	6.89	1 000	1	no ignition	
Nickel 200	34.46	5 000	1	no ignition	
Nickel 200	55.13	8 000	6	no ignition	

[a]In most cases each test represents an average of two values obtained between observation ports 1 and 2 and observation ports 2 and 3.
[b]SD is standard deviation of measured valves.

The effects of *Pox* on $\bar{V}$ for Type 316 stainless steel at *Pox*'s up to 68.91 MPa (10 000 psig) are illustrated in a plot of log $\bar{V}$ versus log *Pox* (Fig. 4). The results indicated that the increase in $\bar{V}$ was proportional to $Pox^{0.075}$ for *Pox*'s between 6.89 and 27.57 MPa (1000 and 4000 psig). At higher *Pox*'s, the dependency of $\bar{V}$ on *Pox* was $\bar{V} \propto Pox^{0.26}$.

Kirschfeld [9] also observed similar changes in the burn rate curve for Type 316 stainless steel wires with diameters of 2 mm (0.076 in.), (Fig. 2). However, he reported that $\bar{V}$ was proportional to the $Pox^{0.5}$.

The effect of *Pox* on $\bar{V}$ for Inconel 718 is illustrated in Fig. 5 as a plot of log $\bar{V}$ versus log *Pox*. The results indicated that the increase in $\bar{V}$ was proportional to $Pox^{0.25}$ over the pressure range between 27.57 and 68.9 MPa (4000 and 10 000 psig).

The effect of *Pox* on $\bar{V}$ for aluminum 6061 is illustrated in Fig. 6 as a plot of log $\bar{V}$ versus log *Pox*. The results indicated that the increase in $\bar{V}$ was proportional to $Pox^{0.58}$ over the pressure range between 6.89 to 68.89 MPa (1000 and 10 000 psig). Kirschfeld [12] reported that the $\bar{V}$ for aluminum (2.0 mm in diameter specimens) decreased as pressure was increased between 10 MPa

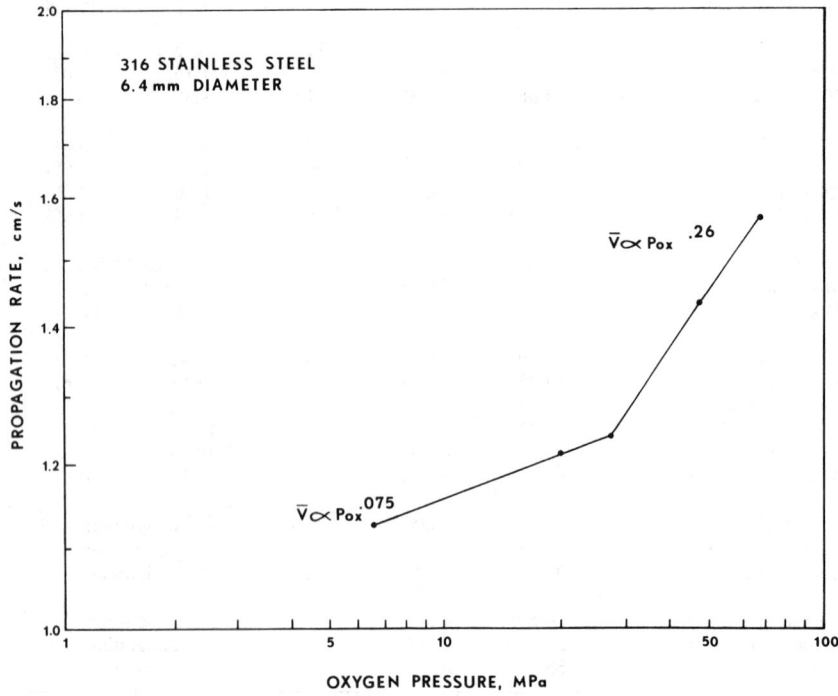

FIG. 4—*Effect of oxygen pressure on the burn propagation rate of Type 316 stainless steel.*

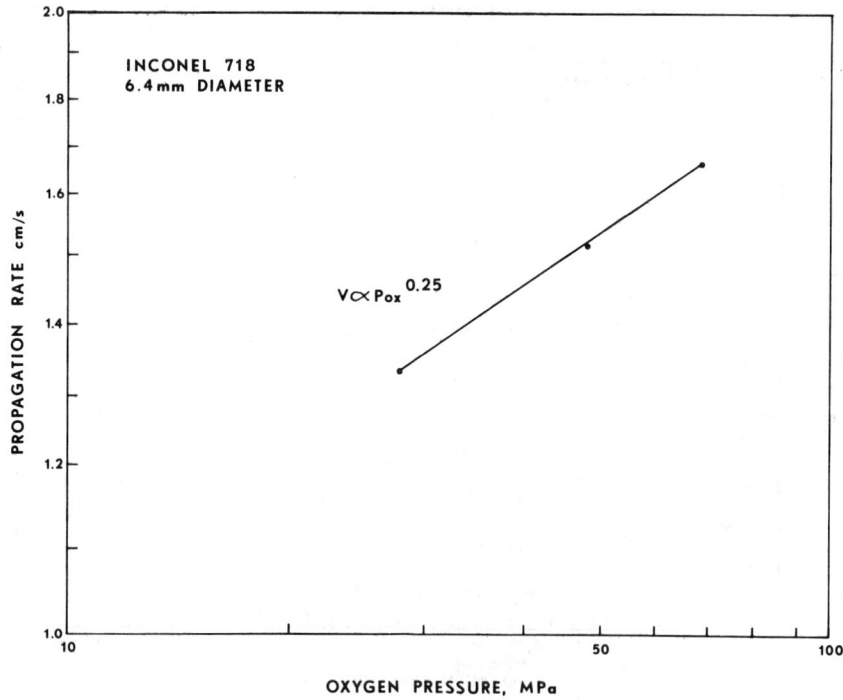

FIG. 5—*Effect of oxygen pressure on the burn propagation rate for Inconel 718.*

(1450 psig) and his maximum test pressure of 20.5 MPa (2975 psig) (Fig. 1). A decrease in $\bar{V}$ as pressure was increased was not supported by the data obtained in this study to date.

Effects of Initial Specimen Temperature

A limited number of tests were conducted in which the test specimens were heated electrically to the ignition temperature of the magnesium promoter (approximately 850 K). The temperature difference along the length of the specimen varied by approximately 111 K (200°F) depending on the test material. The purpose of these tests was to determine if Nickel 200 and Monel 400 could be ignited and propagate the length of the specimen. The results are given in Table 6 for Nickel 200, Monel 400, and Type 316 stainless steel. The results for Nickel 200 and Monel 400 were approximately the same as observed for the ambient temperature tests (Table 3). Nickel 200 did exhibit what appeared to be some burning. The results for Type 316 stainless steel indicated that $\bar{V}$ increased at the higher temperature.

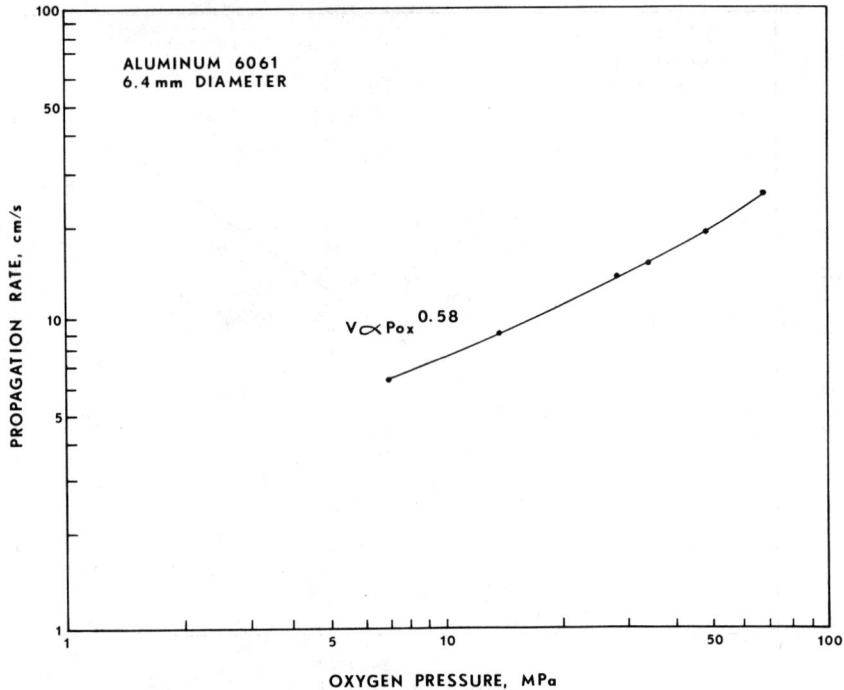

FIG. 6—*Effect of oxygen pressure on burn propagation rate for Aluminum 6061.*

Discussion

Rate Determining Steps for Combustion of Type 316 Stainless Steel and Inconel 718

The effects of Pox on $\bar{V}$ for Type 316 stainless steel and Inconel 718 provide insight as to the rate controlling step in the combustion mechanism. As pointed out in the Background Section, Sato and coworkers [19,21,23] reported that for mild steel the effect of Pox on $\bar{V}$ was attributed to the pressure effect on the combustion rate R_{ox}

$$\bar{V} \propto R_{ox} \qquad (1)$$

Therefore

$$R_{ox} \propto Pox^{\ell} \qquad (3)$$

It is also reasonable to assume that their relationships can be applied to Type 316 stainless steel and Inconel 718. Thus, based on the data illustrated in Figs. 4 and 5, the pressure affect on R_{ox} for Type 316 stainless steel and Inconel 718 can be expressed by following relationships

TABLE 6—*Effects of elevated temperature (approximate 850 K) on ignition and burn propagation.*

| Material | Pressure | | Number of Tests | $\bar{V}$, cm/s (SD)[a] |
	MPa	(psig)		
Nickel 200	6.89	1000	2	no ignition
Nickel 200	27.56	4000	2	no ignition
Monel 400	6.89	1000	2	partial burn[b]
Monel 400	27.57	4000	2	partial burn[b]
316 SS	6.89	1000	3	1.86 (0.03)
316 SS	27.57	4000	3	2.07 (0.04)

[a]Standard deviation.
[b]Less than 10% of the specimen was consumed.

316 stainless steel

$$R_{ox} \propto Pox^{0.075}; \quad 6.89 \text{ to } 27.57 \text{ MPa} \qquad (4)$$
$$(1000 \text{ to } 3000 \text{ psig})$$

$$R_{ox} \propto Pox^{0.26}; \quad 27.57 \text{ to } 68.91 \text{ MPa} \qquad (5)$$
$$(3000 \text{ to } 10\ 000 \text{ psig})$$

Inconel 718

$$R_{ox} \propto Pox^{0.25}; \quad 27.57 \text{ to } 68.91 \text{ MPa} \qquad (6)$$
$$(3000 \text{ to } 10\ 000 \text{ psig})$$

The effects of Pox on oxidation kinetics and associated oxidation mechanisms have been described in several tests [25,27,28] and summarized by Hirano et al for burning of mild steel [19]. The results obtained in this study suggest that the rate determining steps are not physical or chemical adsorption of oxygen on the oxide surface. Physical adsorption is independent of Pox ($\ell = 0$) whereas chemical adsorption is directly dependent on Pox ($\ell = 1$). It would appear that the rate determining steps are either incorporation of oxygen atoms into the oxide layer ($\ell = 0.5$) or ion diffusion through the oxide layer (metal-deficient semiconductor; $\ell = 1/a$ where $a > 0$). However, high speed photography of the burning process indicates that the molten mass is fairly turbulent. How this turbulence affects the combustion process is uncertain at this time and, thus, makes any further conclusions about the rate determining steps purely speculative.

Rate Determining Step for Combustion of Aluminum

The effect of Pox on $\bar{V}$ for aluminum is more complicated than that observed for stainless steel or Inconel. This was evident from the studies by

Kirschfeld [10,16] (Fig. 1) and similar studies by Sato and coworkers [18,19]. Since aluminum burns as a vapor, vaporization of aluminum plays a dominant role in the combustion mechanism. As pointed out in the Background Section, Sato and coworkers reported that the variation in $\bar{V}$ with the increase in Pox was due to the increase in the boiling temperature of aluminum and the decrease in the nonboiling period in the burning process.

Kirschfeld [10] reported that at Pox's above 10 MPa (1450 psig) aluminum exhibited another decrease in $\bar{V}$ as Pox was increased (Fig. 1). He attributed this later decrease in $\bar{V}$ to a decrease in the evaporation rate of aluminum and a decrease in the diffusion properties of the reaction species caused by the increase in Pox. However, in the present study, the effect of Pox on $\bar{V}$ for aluminum was investigated at Pox's three times larger than the maximum Pox's studied by Kirschfeld, and $\bar{V}$ was observed to increase with an increase in Pox (Fig. 6).

Conclusions

Five metals and alloys at ambient temperatures were exposed to a high energy ignition promoter (0.15-g magnesium), and average burn propagation $\bar{V}$ rates were measured for upward propagation in oxygen. Copper 102 and Nickel 200 could not be ignited, and Monel 400 exhibited only partial burning. Of the materials for which $\bar{V}$'s could be measured, Aluminum 6061 exhibited significantly larger $\bar{V}$'s then Inconel 718, 316 stainless steel or 304 stainless steel. Increasing Pox resulted in an increase in $\bar{V}$. The results for Type 316 stainless steel indicated that the dependency of $\bar{V}$ on Pox was $Pox^{0.075}$ for Pox's between 6.89 and 27.57 MPa (1000 and 4000 psig). At higher Pox's, 27.57 to 68.9 MPa, (4000 and 10 000 psig) Type 316 stainless steel exhibited $\bar{V}$'s with a pressure dependency of $\bar{V} \propto Pox^{0.26}$. The effect of Pox, between 6.89 and 68.9 MPa (1000 and 10 000 psig) on $\bar{V}$ for Inconel 718 was observed to be $\bar{V} \propto Pox^{0.25}$. The pressure dependency of $\bar{V}$ for aluminum 6061 was observed to be $\bar{V} \propto Pox^{0.58}$ for Pox's between 6.89 and 68.9 MPa (1000 and 10 000 psig).

Increasing the temperature of the test specimen from ambient to approximately 850 K (1070°F) has little affect on the ignition characteristics of Nickel 200 or Monel 400. However, in the case of Type 316 stainless steel, an increase in $\bar{V}$ was observed at this higher temperature.

Acknowledgments

The authors wish to acknowledge Carl Wright, Leo Hall, Mike Rundell, Ron Gamboa who conducted the majority of the tests from which the data presented in this paper was obtained. The funds for this work have been provided by the Office of the Chief Engineer, NASA Headquarters, Washington, DC, primarily through the efforts of Joyce McDevitt.

References

[1] Lapin, A., "Liquid and Gaseous Oxygen Safety Review," NASA-CR-120922, Vol. I-IV, APCI TM184, National Aeronautics and Space Administration Lewis Research Center, Cleveland, OH, June 1972.

[2] Yuen, H. H., "Compatibility of Materials with Oxygen NAEC Report: NAEC Misc. 92-0354," Naval Air Engineering Center, Lakehurst, NJ, Sept. 1978.

[3] Schmidt, H. W. and Forney, D. F., "ASRDI Oxygen Technology Survey, Volume IX: Oxygen Systems Engineering Review," National Aeronautics and Space Administration SP-3090, National Aeronautics and Space Administration, Washington, DC, 1975.

[4] Clark, A. F. and Hust, T. G., "A Review of the Compatibility of Structural Materials with Oxygen," *AIAA Journal,* Vol. 12, No. 4, April 1974, pp. 441-454.

[5] "Oxygen Fire Event Report Fire Laboratories Test Office," NASA White Sands Test Facility, Las Cruces, NM.

[6] Wegener, W., "Investigations on the Safe Flow Velocity to be Admitted for Oxygen in Steel Pipe Lines, " *Stahl und Eisen,* Vol. 84, No. 8, 1964, pp. 469-475.

[7] Benz, F. J., Williams, R. E., and Armstrong, D., "Ignition of Metals and Alloys by Impact of High Velocity Particle," paper presented at the Symposium on Flammability and Sensitivity of Materials in Oxygen-Enriched Atmosphere, Washington, DC, April 1985.

[8] Kirschfeld, L., "Influence of Geometric Form on the Combustion Rate of Sheet Metal in Oxygen, "*Archiv fur den Eisenhuttenw,* Vol. 33, Sept. 1962, pp. 617-621.

[9] Kirschfeld, L., "Combustibility of Steel and Cast Iron in Oxygen at Pressures of Up to 150 Atmosphere," *Archiv fur den Eisenhuttenw,* Vol. 39, July 1968, pp. 535-539.

[10] Kirschfeld, L., "Apparatus for Combustion Tests on Metals Under Oxygen Pressures Up to 200 Atmospheres" and "The Combustibility of Iron Wire in High Pressure Oxygen," *Archiv fur den Eisenhuttenw,* Vol. 36, Nov. 1965, pp. 823-826.

[11] Kirschfeld, L., "Combustibility of Metals in Oxygen," *Angew. Chem.,* Vol. 71, 1959, pp. 663-667.

[12] Kirschfeld, L., "Combustibility of Metals in Oxygen Up to 200 Atmospheres Pressure," *Metallurgy,* Vol. 21, No. 2, Feb. 1967, pp. 98-102.

[13] Kirschfeld, L., "The Combustion Rate of Light Metal Wires in Oxygen," *Metallurgy,* Vol. 14, 1960, pp. 213-219.

[14] Kirschfeld, L., "Combustion Rates of Non-Iron Metal Wires in Oxygen, *Metallurgy,* Aug. 1960, pp. 792-796.

[15] Kirschfeld, L., "Die Verbrennungsgeschwindigkeit von Eisendrahten in Sauerstoff hohen Druckes," *Archiv fur das Eisenhuttenwesen,* Vol. 32, No. 1, Jan. 1961, pp. 57-62.

[16] Kirschfeld, L., "Uber die Verbrennungsgeschwindigkeit von Leichtmetalldrahten in Saverstoff hohen Druckes, *Metallurgy.* Vol. 15, No. 9, Sept. 1961, pp. 873-878.

[17] Harrison, P. L. and Yoffe, A. D., "The Burning of Metals," *Proceedings of Royal Society,* Vol. 261A, 1961, pp. 357-370.

[18] Sato, K., Hirano, T., and Sato, J., "Behavior of Fires Spreading Over Structural Metal Pieces in High Pressure Oxygen," *ASME JSME Thermal Engineering Joint Conference Proceedings,* Vol. 4, 1983, pp. 311-316.

[19] Hirano, T., Sato, K., and Sato, J., "An Analysis of Upward Fire Spread Along Metal Pieces," *Fire Dynamics and Heat Transfer - HTD,* Vol. 25, American Society of Mechanical Engineers, New York, 1983, pp. 83-87.

[20] Hirano, T., Sato, K., Sato, Y., and Sato, J., "Prediction of Metal Fire Spread in High Pressure Oxygen," *Combustion Science and Technology.* Vol. 32, 1983, pp. 137-159.

[21] Sato, J., Sato, K., and Hirano, T., "Fire Spread Mechanisms Along Steel Cylinders in High Pressure Oxygen," *Combustion and Flame.* Vol. 51, 1983, pp. 279-287.

[22] Sato, K., Sato, Y., Tsuno, T., Nakamura, Y., Hirano, T., and Sato, J., "Metal Combustion in High Pressure Oxygen Atmosphere," *Proceedings of the 15th International Congress on High Speed Photography/Photonics,* SPIE, Vol. 348, San Diego, CA, Aug. 1982, pp. 828-832.

[23] Hirano, T., Sato, Y., Sato, K., and Sato, J., "The Rate Determining Process of Iron Oxidation at Combustion in High Pressure Oxygen," *Oxidation Communications,* Vol. 6, Nos. 1-4, 1984, pp. 113-124.

[24] Stoltzfus, J. and Benz, F. J., "Development of Methods and Procedures for Determining

the Ignitability of Metals in Oxygen," NASA Interim Report: TR-281-001 INT-1, White Sands Test Facility, Las Cruces, NM, Nov. 1984.

[25] Kofstad, P., *High Temperature Oxidation of Metals,* Wiley, New York, 1966.

[26] Grosse, A. V. and Conway, J. B., "Combustion of Metals in Oxygen," *Industrial and Engineering Chemistry,* Vol. 50, No. 4, April 1958, pp. 663–672.

[27] Hauffe, K., *Oxidation of Metals,* Plenum Press, New York, 1965.

[28] Kubaschewski, O. and Hopkins, B. E., *Oxidation of Metals and Alloys,* 2nd ed., Butterworth and Co., London, 1967.

Michael A. Benning[1] and Barry L. Werley[1]

The Flammability of Carbon Steel as Determined By Pressurized Oxygen Index Measurements

REFERENCE: Benning, M. A. and Werley, B. L., **"The Flammability of Carbon Steel as Determined by Pressurized Oxygen Index Measurements,"** *Flammability and Sensitivity of Materials in Oxygen-Enriched Atmospheres: Second Volume, ASTM STP 910*, M. A. Benning, Ed., American Society for Testing and Materials, Philadelphia, 1986, pp. 153-170.

ABSTRACT: The ASTM oxygen index, a standard atmospheric pressure test for comparing the flammability of plastics, has been used in combination with other tests to evaluate the suitability of nonmetals for use in oxygen service. In 1983, this test was extended to allow measurements at pressures to 2 MPa (300 psig). This paper describes how the concept of a pressurized oxygen index test can be extended further to measure the flammability of metals and hence their suitability for use in oxygen service. An experimental program is described that measured oxygen indexes for carbon steel pipe (25 mm [1.0 in.] diameter by 4.8 mm [0.188 in.] wall) as a function of pressure to 20 MPa (3000 psig) at temperatures of about 300 and 355 K (80 and 180°F). Oxygen/nitrogen mixture velocities inside the pipe varied from 0.15 to 0.90 m/s (0.5 to 3 ft/s). An experimental system was developed, including the test specimen configuration and a thermite ignition pill for use in oxygen mixtures. Measured oxygen indexes were 81 mol% oxygen in nitrogen at 1.03 MPa (150 psi), 53 mol% at 6.9 MPa (1000 psi), and 51 mol% at 20.7 MPa (3000 psi). Additional tests demonstrated the effectiveness of Monel as a firebreak material and Buna N rubber as an ignition source for carbon steel.

KEY WORDS: flammability (metals), carbon steel, ignition, fires, hazards, safety, oxygen index, Monel

In 1978, Air Products and Chemicals, Inc. became interested in a tertiary oil recovery process in which air is pumped down carbon steel well bores to burn underground residual oil. The heat generated reduces the viscosity of the oil ahead of the combustion zone, allowing it to flow to surrounding recovery wells. The oil industry has investigated ways to improve this technique by

[1]Lead engineer and hazard research specialist, respectively, Air Products and Chemicals, Inc., P.O. Box 538, Allentown, PA 18105.

substituting oxygen-enriched air or commercially pure oxygen for the air. Although this change was shown to be economically and technically advantageous [1], the combination of typically high operating pressures of 7 to 20 MPa (1000 to 3000 psi)[2] and the oxygen could result in ignition of the carbon steel well bore piping and destruction of the well.

To evaluate the risks of oxygen enrichment, an experimental program was conducted to measure the oxygen index of carbon steel as a function of pressure. This allowed determination of threshold oxygen concentrations below which the steel should not burn even if ignited, thus minimizing the risk. This approach also allowed some assessment of the risk at concentrations above the index. Measurements were also made to evaluate the use of Monel as a firebreak at conditions above the oxygen index, to limit the extent of damage should the steel ignite.

The oxygen index, which is a measure of the threshold oxygen concentration needed for flame propagation, has been established by ASTM as a standard measurement for comparing the flammability of plastics (ASTM Method for Measuring the Minimum Oxygen Concentration to Support Candle-like Combustion of Plastics [Oxygen Index] [D 2863]). The standard test is conducted at atmospheric pressure, but subsequent modifications by Benning [2] allow measurements at pressures up to 2 MPa (300 psi). The standard test is normally conducted at uncontrolled ambient temperatures. However, increased temperature reduces the oxygen index as shown for nonmetals by Hendrix et al [3] and for steel by this work.

This paper reports the results of applying the principles of the oxygen index test to measure the flammability of metals at pressures up to 20 MPa (3000 psi). However, the configuration used deviated substantially from the standard oxygen index test because key parameters related to the flow of oxidant in well bore piping had to be simulated.

Few data were available on metal flammability in gaseous mixtures containing between 21 and 100% oxygen [4–7]. Furthermore, most data were for fine wires, ribbons, and similar materials, which do not necessarily reflect the performance of large-scale systems. Based on knowledge of the oxygen index test, it was known that a combustible material has an oxygen threshold concentration below which it will not burn, even if ignited. However, flame propagation is not a physical property, it is a behavioral property. Hence, the threshold concentration for flame propagation will depend on numerous factors, including geometry, pressure, temperature, and degree of heat sink [8]. Furthermore, the dependence of the threshold on these factors may vary from material to material.

This program began with the design and construction of a vessel to overcome the problem of containing complete combustion of the fairly large specimens selected at pressures to 20 MPa (3000 psi). Also, an ignition source had

---

[2]Pressures throughout are gage pressures.

to be chosen that would provide ignition events of comparable effectiveness in gases of varying oxygen concentration. A thermite reaction was chosen to fulfill this requirement.

The program was designed to simulate wall thicknesses of carbon steel well bore piping and typical gas velocities and well bore temperatures, and also to explore upward rather than downward propagation mechanisms (since this would be the most probable direction in a well bore). Following identification of the threshold concentration, materials that might be present in well bores (namely, Buna N or crude oil) were briefly studied to see if they could ignite carbon steel near the threshold concentration.

Once the experimental system was devised, specific experiments were conducted to accomplish the following objectives:

• Measure the oxygen index, or threshold concentration, of commercially pure oxygen (about 99.7% oxygen and 0.3% argon) in nitrogen that will just support upward flame propagation of strongly ignited carbon steel in slowly downward moving gas streams.

• Measure the change in oxygen index in the range 1 to 20 MPa (150 to 3000 psi).

• Determine whether materials that may be present in typical well bores (such as Buna N or crude oil), if ignited, are capable of igniting carbon steel.

• Determine whether Monel 400 could be used as an effective firebreak at conditions above the oxygen index, to limit the amount of damage should the steel ignite.

Experimental System

Ignition Pill Development

In the past [9–11], metal ignition work frequently used "ignition pills" consisting of flammable metals, usually a mixture of steel wool and aluminum wool, which was then ignited by an electric match. The heat of combustion released as these components burned in an oxygen environment provided ignition energy to the test specimen.

In this program, the use of oxygen and nitrogen mixtures was felt to invalidate the use of these pills, since they rely on the ambient oxygen for their combustion. The intensity of metal fires is well known to decrease with decreases in ambient oxygen concentration. Furthermore, specimen ignition may fail at a lower oxygen concentration for more than one reason: the specimen could be below its oxygen index; the ignition pill could become ineffective because of reduced intensity of its own combustion process; or the pill could even be below its own oxygen index.

Because of these problems, an alternate ignition method that would be independent of oxygen concentration was sought. Because of the uncertainty of the effect of oxygen dilution, even very flammable ignition materials, such as magnesium, which are known to burn in low oxygen concentrations, were rejected for the same reasons the conventional pills were rejected. In addition, Joule heating was considered to be too complex, laser ignition too complex and costly, and arc/spark too difficult to achieve at 20 MPa. Ultimately, a thermite reaction was selected for a new ignition pill design.

Thermite provides a method to supply a reasonably constant ignition energy in any atmosphere. The thermite reaction is shown below

$$Fe_2O_3 + 2Al \rightarrow Al_2O_3 + 2Fe$$

It produces a heat release regardless of the ambient oxygen concentration. The heat release is not, however, perfectly constant because reoxidation of the reduced iron oxide may vary in differing oxygen concentrations or differing mechanical configurations. This reoxidation of the iron, however, produces at most 46% of the total heat release for a stoichiometric mixture.

The thermite pill used experimentally was prepared from an approximately stoichiometric mixture of aluminum filings (1 g) and commercial iron oxide powder (3 g). From published heats of combustion for aluminum (31.0 kJ/g) and iron (7.4 kJ/g) [12], one can predict a 31.0-kJ heat release if the pill was reacted in oxygen and complete reoxidation of the iron occurred. If the pill were reacted in nitrogen so that no iron reoxidation was possible, one would predict a 15.5-kJ release. Experimental measurement of the heat release in nitrogen yielded a value of about 16.7 kJ, which was in reasonable agreement with the predicted value. However, the measured value in oxygen at 0.2 MPa (30 psi) was only 23 kJ, which indicated that only partial reoxidation of the iron occurred in the calorimeter. The degree of reoxidation throughout the test program at varying pressures and concentrations probably differed from this measured value, but no attempt was made to quantify it.

The final pill configuration was selected after a preliminary sequence of tests with various materials of inadequate properties. Aluminum powder (Baker 0446) mixed with ferric oxide and pressed dry into a pellet was quite explosive in bench tests and did not release its heat in the required local fashion. Furthermore, the pill crumbled easily. However, when water was added to the components to form a paste that could be pressed into a stronger pill, a spontaneous, exothermic reaction occurred during drying. The heat generation was attributed to reaction between the aluminum powder and the water. This reaction apparently generated steam pressure internally, which burst the pill. The heat release was not caused by the thermite reaction.

Finally, relatively coarse 6061-T6 aluminum filings, generated by hand filing, were substituted for the aluminum powder. Structurally sound pills could now be formed, but they could not be directly ignited by electric match or by an electrically heated wire. Therefore, they were indirectly ignited by using an electrically heated wire immersed in a thin layer of thermite powder on top of the pill that had been prepared with aluminum powder rather than aluminum filings (see Fig. 1).

These pills usually reacted predictably. That is, upon ignition they coalesced into a single molten mass and released their heat into a single location.

A few tests were conducted with carbon steel plates to select the mass of thermite in the final pill. A central core of 4 g of thermite successfully ignited a 2.4-mm-thick carbon steel disk in roughly 1.7-MPa oxygen, and, hence, was chosen for the test program. A pill using 2 g of thermite did not ignite a 2.4-mm-thick disk.

Specimen Development

A tubular specimen with a closed bottom to support the ignition pill was chosen initially (Fig. 2.1). The closed end was countersunk, so that its minimum thickness just below the pill was 2.0 mm (0.080 in.) at the site of the 1.7 mm (0.67-in.) diameter center hole. These dimensions were selected because the ignition pill had successfully ignited an even thicker 2.4-mm carbon steel disk, containing a similar hole, in 1.7-MPa oxygen. However, at 6.9 MPa, the specimen did not ignite, even in 99.7% oxygen. It was concluded that the pill had reacted and melted, and was then blown through the hole before sufficient heat had been transferred to the specimen for ignition to occur.

The specimen was modified (Fig. 2.2) by peening a plug of silver solder into a larger diameter hole. Also, two purge holes of 0.34 mm (0.0135 in.) diameter were drilled just above the pill. It was anticipated that the bottom of the

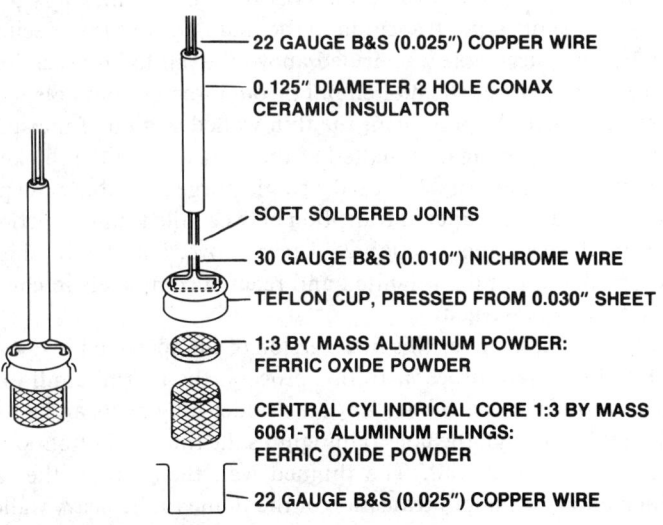

FIG. 1—*Thermite pill assembly.*

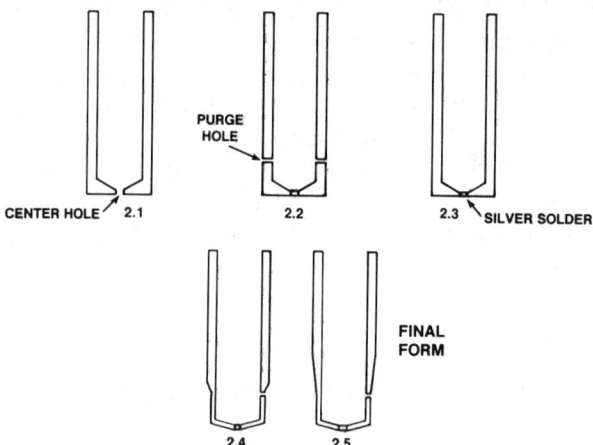

FIG. 2—*Evolution of test specimens.*

specimen would be preheated enough to melt the solder (about 920 K, 1200°F), thus opening the bottom hole and allowing the molten pill products to pass through and ignite the specimen. This modification did produce complete reaction in 99.7% oxygen, but it produced negligible and, therefore, inconclusive damage at 80%. A second modification (Fig. 2.3), eliminating the purge holes, was made to increase the pressure drop across the solder plug. However, this yielded inconclusive damage. In a third modification (Fig. 2.4), the bottom 18 mm (0.70 in.) was reduced in thickness to 2 mm (0.080 in.) with a short taper transition to the main body of the specimen. In addition, only one purge hole was drilled above the pill to increase the available pressure drop to force the molten pill through the bottom hole when the silver solder melted. In 80% oxygen, the thin-walled section of the specimen was combusted, but propagation halted at the transition to the thicker wall.

The final modification (Fig. 2.5) used a single purge hole above the pill and was prepared with a long, tapered transition between the thinned section near the pill and the heavy walled section 25.4 mm above. Ultimately, this specimen, shown in detail in Fig. 3, ignited and reacted completely in concentrations as low as 47.2% oxygen.

The ignition mechanism for this fifth version of the specimen is viewed as a multistep kindling chain process. In this process, the thermite pill preheats the thinned bottom of the specimen until the solder plug melts and allows the molten pill to flow through the hole. This ignites the thinned bottom which, in turn, ignites the thinned wall. The thinned wall then ignites the tapered length, which allows a smooth transition of the flame to the heavy walled test section.

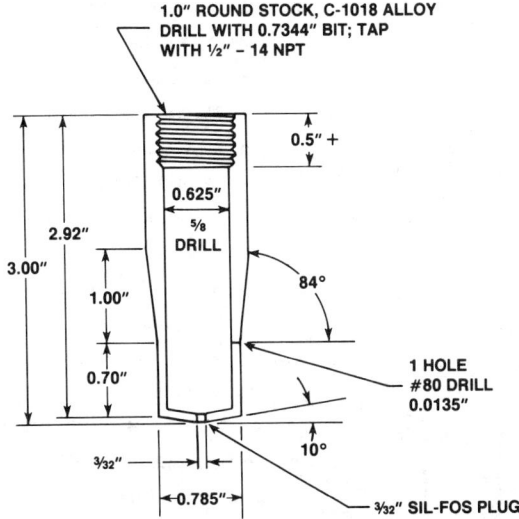

NOTE: SPECIMEN IS DRILLED WITH ⅝" BIT HAVING A 160° POINT ANGLE

FIG. 3—*Details of final specimen configuration.*

Test Vessel Description

The tests were conducted in a high-pressure vessel designed to contain the reaction. Components of the vessel are designated by letters on Fig. 4, and these letters are referred to in the discussion below.

The vessel was fabricated from 158-mm (6¼ in.) diameter stainless steel end plates that were about 27 mm (1.06 in.) thick (Items c and p); a 160-mm (6.31-in.) long section of 4-in. double extra-heavy wall (17.1 mm [0.674 in.]) ASTM-B43 red brass pipe (Item f); and eight 15.9-mm (⅝-in.) diameter American Society of Mechanical Engineers (ASME) SA-193-B7 alloy national coarse threaded rods (Item l, typical). Matching shallow ridges were machined in the end plates and pipe to allow centering of the pipe. 0-ring grooves in the pipe housed Circle Seal 242 Viton 0 rings. To achieve the desired 20 MPa pressure capability, the clearance between the threaded rods and brass pipe was limited to about 2 mm (this minimized bowing of the end plates due to initial bolt loading).

Brass disks (Items d and o), 6.4 mm (¼ in.) thick, were used to shield the stainless steel end plates from reaction spatter. These disks were closely fitted to the inside diameter of the brass pipe. Copper exhaust gas vents were fed through the top stainless steel end plate and the brass disk shield so that there would be no direct contact of the hot exhaust gases with the end plate.

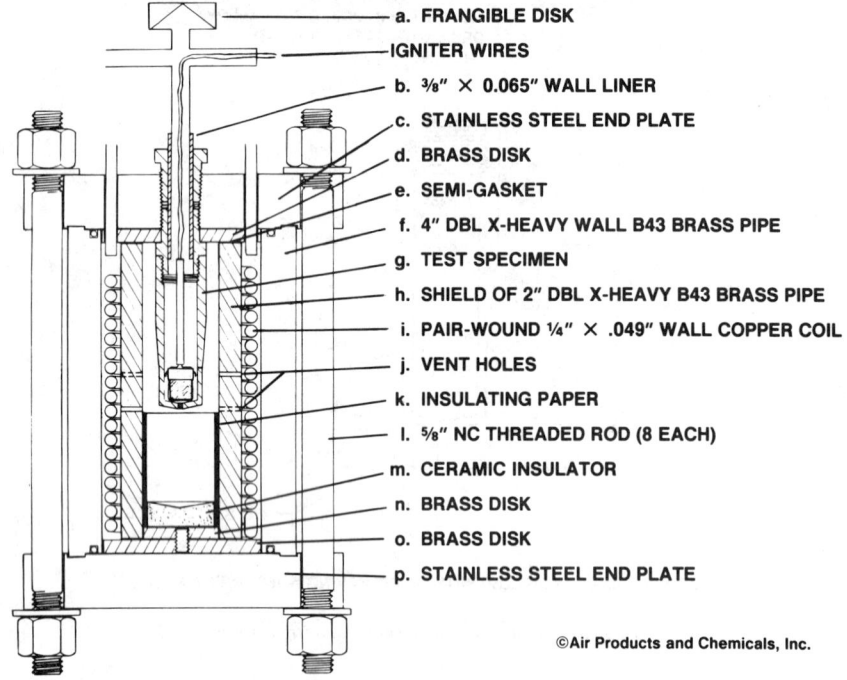

a. FRANGIBLE DISK

IGNITER WIRES

b. ⅜" × 0.065" WALL LINER

c. STAINLESS STEEL END PLATE

d. BRASS DISK

e. SEMI-GASKET

f. 4" DBL X-HEAVY WALL B43 BRASS PIPE

g. TEST SPECIMEN

h. SHIELD OF 2" DBL X-HEAVY B43 BRASS PIPE

i. PAIR-WOUND ¼" × .049" WALL COPPER COIL

j. VENT HOLES

k. INSULATING PAPER

l. ⅝" NC THREADED ROD (8 EACH)

m. CERAMIC INSULATOR

n. BRASS DISK

o. BRASS DISK

p. STAINLESS STEEL END PLATE

©Air Products and Chemicals, Inc.

FIG. 4—*High-pressure test vessel.*

A cylindrical shield (Item h) of 2 in. double extra-heavy wall (11.1-mm [0.436-in.]) ASTM-B43 brass pipe was used to contain radial spatter from the specimen. It also served to reduce the water volume of the vessel and thereby minimized the immediate oxidant availability in the event of a violent reaction. Through the wall of the shield were 16 vent holes (Item j), 2 mm (0.08 in.) in diameter, uniformly spaced in two rows both at the level, and 16 mm above the level, of the bottom of the specimen. These holes served to channel the hot combustion gases to a cooling coil in the annulus between the inner shield and outer vessel wall. The cylindrical shield was centered on a 6.4-mm (¼-in.) thick brass disk (Item n) of close fit to its inside diameter. This disk was attached by a threaded, copper stud to the larger brass disk that shielded the lower end plate. A semigasket (Item e) of Cotronics 390-40-3 insulating paper was placed between the top of the cylindrical shield and the upper end plate shield. This semigasket reduced the possibility of exhaust gas flow up along the outside of the specimen, which might have promoted combustion through a preheating effect. After a run, the observed "crush" upon removal of the semigasket indicated proper assembly tolerances had been achieved.

The entire length of the annulus between the shield and outer pipe was taken up by a pair-wound helical coil of copper tubing (Item i). Water was circulated through this tubing to remove and redistribute the specimen's heat of combustion before it could weaken the vessel wall.

Within the shield were located two insulators. A liner (Item k) of Cotronics 390-40-3 insulating paper to the level of the lower vent holes was used to slow the transfer of heat to the shield and to facilitate removal of the slag when cooled. A 13-mm-thick disk of Cotronics RTC-60 castable ceramic (Item m) was used to catch the slag and slow the transfer of heat to the vessel bottom. The volume of the space above the ceramic insulator and below the level of the vent holes was chosen to exceed the converted slag volume so that vent hole plugging would be minimized.

A test specimen (Item g), with ignition pill installed, was mounted on a shortened, brass hex-close nipple, which had its lower thread modified to be parallel instead of tapered. A liner (Item b) of 9.5-mm (^{3}/8-in.) diameter copper tubing extended almost to the end of this nipple and passed through both the end plate and the male connector above. This liner shielded the end plate from direct contact with hot gases if the reaction backed up or if the rupture disk on the inlet stream (Item a) released. The test gas mixture and igniter wires passed through this liner tube.

Apparatus and Procedure

As shown on the flow sheet for the experimental apparatus in Fig. 5, most of the equipment was located within a steel barricade with valves operated remotely from outside. To begin a test, the specimen and vessel were purged with the oxidant gas mixtures. The cooling water supply system was pressurized and flow through the internal cooling coils was begun. Instrumentation and temperature recorders were started.

The system was pressurized and the gas mixture flow adjusted to the correct value. Ignition was effected by applying power to the igniter wire in the thermite powder. When thermal activity on the exhaust gas temperature recorder became insignificant (that is, it either achieved its maximum or it failed to respond), the gas flow was halted, the vessel vented, and the remaining cooling water supply allowed to deplete itself.

The test gas mixture was supplied to pressure regulator R, then, at the proper pressure, it went to shutoff valve V1, through two check valves to a vent valve and through the test vessel, through an exhaust gas cooling coil, to three parallel sintered filters, and through flow adjusting valve V2. It was finally vented through a rotameter. The test vessel was submerged in water as an added precaution against localized heating.

The cooling water circuit was pressurized with nitrogen through a cylinder regulator adjusted to 1.1 MPa (160 psi); this pressure was chosen to prevent flash evaporation and even momentary interruption of the cooling water flow. The cooling water flowed from the supply vessel through a shutoff valve with an extended stem to allow operation from outside the barricade. It then

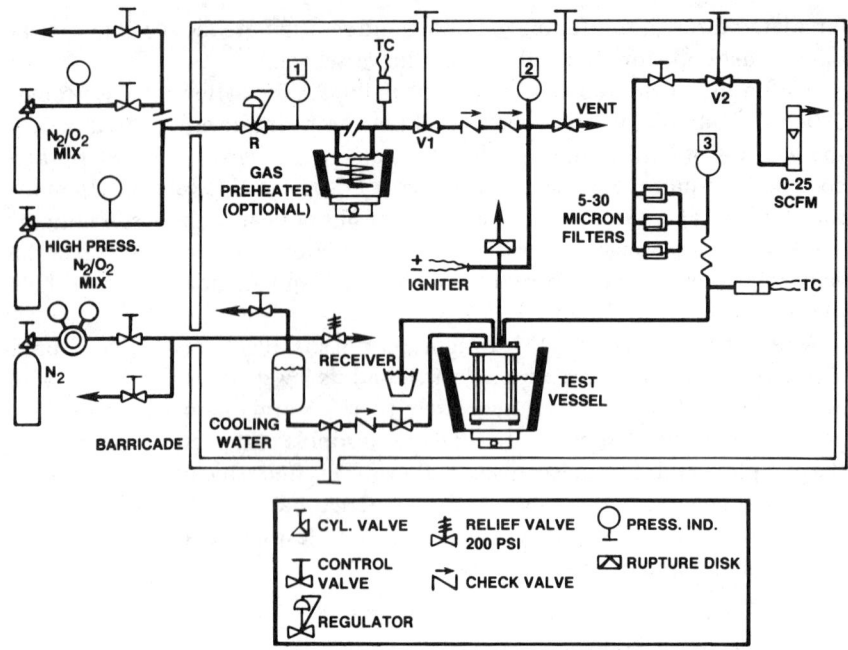

FIG. 5—*Flowsheet for apparatus.*

flowed through a check valve, and a valve inside the barricade preset to establish a flow rate of 1 L/min. It then flowed to the internal cooling coils, and was finally vented into a receiver.

The test gas composition was measured with a Servomex paramagnetic oxygen analyzer that had been calibrated against commercially pure oxygen (about 99.7% oxygen, 0.3% argon) and nitrogen (for the 100% and 0% of scale points). This analyzer could be read to about 0.1 percentage point. Since many gas mixtures were prepared in the laboratory, they were analyzed before and after each test to confirm that there was no stratification within the mixture. In one case a 0.2 percentage point variance (attributed to faulty analysis rather than stratification) was observed; however, in all other cases the analysis was reproduced to within 0.1 percentage point.

Tests at 355 K (180°F)

In those tests conducted at 355 K, the feed gas mixture was passed through a preheater of copper tubing located in a hot water bath. In addition, the water bath surrounding the vessel was heated and the cooling water vessel was filled with water at 355 K. The cooling water vessel and interconnecting lines were insulated. Before the run, hot water was circulated through the cooling water vessel, the test vessel, the test vessel bath, and the test gas preheater bath to bring the entire system to temperature. A thermocouple was located in the discharge line of the test gas preheater to measure and set the temperature of the gas before ignition.

Discussion of Results

Even Above the Oxygen Index, Ignition May Fail

When gas concentrations are plotted to show the limits where combustion does and does not occur (flammability maps), the reliability of the nonflammable regions is always subject to a degree of uncertainty. When propagation does occur, the conditions can be considered in the flammable region. In contrast, the absence of propagation does not guarantee that those test conditions are outside of the flammable region. Lack of propagation may be due to nonflammability, but it may also be due to spurious extinguishment, inadequate ignition energy, inadequate ignition temperature, or a "misfire."

As a result, the following criteria were established to appropriately evaluate the test results:

• Complete, or at least extensive, combustion was positive proof that the test condition was within the flammable region. Complete combustion indicated that the oxygen concentration was equal to or greater than the oxygen index.

• Partial combustion was a strong indicator that ignition was acceptable, and that the test concentration was below the oxygen index.

• Nonignition was not conclusive.

At each particular pressure studied, the objective was to find the lowest oxygen concentration that would allow propagation and the highest concentration that would preclude propagation. The higher oxygen concentration was then used as the oxygen index.

Carbon steel is difficult to ignite. In this and other test programs, metals have often been exposed to ignition events of considerable energy without reacting significantly. As discussed earlier, following development of an acceptable ignition pill, the specimen configuration had to be modified several times before the carbon steel would ignite and burn extensively.

Oxygen Index Data at Room Temperature

A typical set of test specimens that burned at 6.9 MPa (1000 psi) is shown in Fig. 6. The specimens (1 through 5) underwent partial combustion at oxygen concentrations up to 51%, including two specimens (4 and 5) at gas velocities different from the normal 0.3 m/s. At 53% oxygen, the specimen (6) was completely consumed. This value, therefore, was taken as the oxygen index, although additional testing at concentrations between 51 and 53% could have

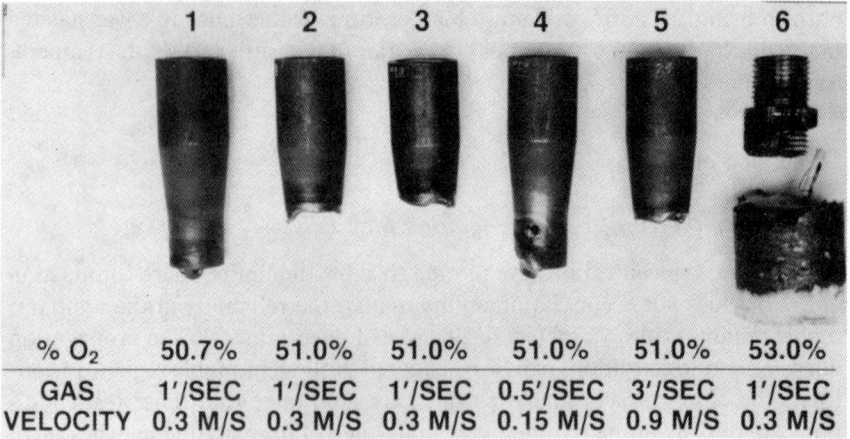

% O₂	50.7%	51.0%	51.0%	51.0%	51.0%	53.0%
GAS VELOCITY	1'/SEC 0.3 M/S	1'/SEC 0.3 M/S	1'/SEC 0.3 M/S	0.5'/SEC 0.15 M/S	3'/SEC 0.9 M/S	1'/SEC 0.3 M/S

FIG. 6—*Carbon steel specimens after ignition at 6.9 MPa, 300 K.*

pinpointed the threshold concentration more precisely. As a caution, one should note that the measured oxygen index is a concentration equal to or above the minimum oxygen concentration required to support combustion. Allowance should be made for this if one is attempting to define a concentration that will preclude combustion.

The data for all tests with the standard specimen at ambient temperature are shown in Table 1 and plotted in Fig. 7. The graph shows that the index fell rapidly as the pressure increased—from 81% oxygen at 1 MPa (150 psi) to 53% at 6.9 MPa (1000 psi). However, from 6.9 to 20 MPa (3000 psi), the index decreased by only 2 percentage points to 51%.

Effect of Increased Ambient Temperature

In one set of experiments at 20.7 MPa, the temperatures of the specimen and feed gas were raised to 355 K (180°F), to simulate the higher temperature at the bottom of a well bore. There were two partial combustions at 46.5% oxygen and complete combustions at 47.1 and 48.2%. Taking the oxygen index as 47%, this was a reduction of about 4 percentage points from the room temperature index of 51%. However, this latter value is probably on the high side, so the actual reduction is probably less than 4 percentage points.

Effect of Oxidant Gas Velocity

In two tests at 6.9 MPa, the oxidant gas velocity through the specimen was changed from the standard 0.3 m/s. In one test it was half the standard, and

TABLE 1—*Combustion results for C-1018 carbon steel.*[a]

| Gage pressure | | O₂ Concentration, | |
MPa	psi	mol%	Result
1.03	150	56.7	no ignition
		56.8	no ignition
		64.5	no ignition
		79.2	slight combustion
		79.2	partial combustion
		80.9	no ignition, misfire
		80.9	complete combustion, OI[b]
		82.2	complete combustion
		84.2	complete combustion
2.1	300	65.0	slight combustion
2.4	350	65.0	slight combustion
		65.0	complete combustion
2.8	400	64.6	partial combustion
		64.6	complete combustion
3.1	450	64.6	complete combustion
6.9	1000	50.7	slight combustion
		51.0	partial combustion
		51.0	partial combustion
		53.0	complete combustion, OI
		55.3	complete combustion
		56.8	partial combustion, misfire
		60.0	complete combustion
		63.0	complete combustion
		79.2	complete combustion
12.4	1800	48.5	partial combustion
20.7	3000	48.2	partial combustion
		51.0	complete combustion, OI
		53.0	no ignition, misfire
		53.1	complete combustion
		79.2	complete combustion

[a]Gas velocity through specimen 0.3 m/s; room temperature (about 300 K); ignition by 4 g of thermite.
[b]Oxygen index.

in the other it was three times the standard. The oxygen concentration was 51%, just below the measured index at 6.9 MPa.

Photographs of the specimens from these two tests are also shown in Fig. 6. At the higher gas velocity of 0.9 m/s, the damage resulting from ignition was essentially the same as that in the two tests at 0.3 m/s. At 0.15 m/s, the damage was less, but still sufficient to conclude that ignition occurred. These two tests indicate that gas velocity in the range 0.15 to 0.9 m/s has only a small effect on the oxygen index. In particular, these tests indicate that the oxygen index is not significantly reduced at either higher or lower gas velocities. Hence, the values for 0.3 m/s are acceptable from a safety standpoint.

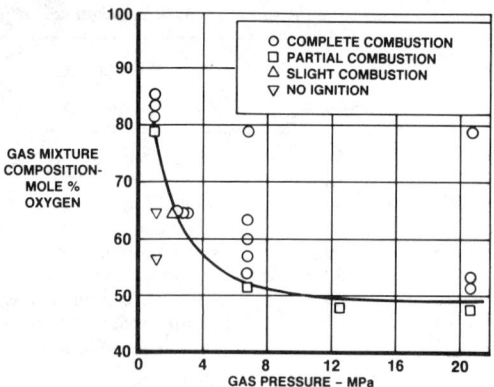

FIG. 7—*C-1018 carbon steel combustion results.*

Ignition with Buna N Rubber

A carbon steel system that is operated at or above its oxygen index must be examined to determine whether an ignition source capable of igniting the steel is likely to be present. Sources include nonmetallic system components or contamination.

This problem was examined experimentally by burning 0.95 g of Buna N rubber in place of the thermite pill. The heat of combustion of the Buna N was about twice that of the pill. The carbon steel specimen ignited at an oxygen concentration of 53.2% at 6.9 MPa. This concentration was within 0.2 percentage point of the index, and proves the effectiveness of Buna N as an igniter.

However, achieving ignition was not easy. Several tests in various configurations with both Buna N fuel and crude oil fuel did not ignite the carbon steel. However, these test results were attributed to the mechanics of the ignition process rather than to the materials' inability to ignite the carbon steel.[3]

The initial Buna N ignition test, conducted in 79.1% oxygen, used a quantity of Buna N calculated to release heat equivalent to a thermite pill. When the Buna N, in the form of several 0 rings on an electric match, was substituted for the thermite pill, it did not ignite the metal or even produce the anticipated level of surface oxide. It was concluded that the Buna N was burning in the vapor phase, and that the heat was being carried away by the gas rather than being transferred to the specimen. In contrast, in the thermite pill experiments, once the pill was molten, it was in excellent condition for heat transfer to the ignition zone of the specimen. To resolve the problem in the Buna N tests, a silver solder plug was removed from the bottom of the normal

---

[3]The sequence of steps taken, before ignition occurred with Buna N, was similar to the different configurations that had to be developed before the thermite pills would successfully cause ignition. These results point out again the complexity of achieving metal ignition.

specimen. That way, all combustion products would pass through a 2.4-mm-diameter hole and preheat the thin bottom of the specimen to its ignition temperature. This resulted in ignition of the specimen, in both 79.0 and 53.2% oxygen.

The Effectiveness of Monel as a Firebreak

This experimental program also evaluated the effectiveness of Monel as a firebreak material. Direct ignition of Monel specimens in commercially pure oxygen at 6.9 and 13.8 MPa (1000 and 2000 psi) was attempted using procedures identical to those employed with the carbon steel specimens. Figure 8 shows that ignition was successful only to the extent that significant damage resulted; propagation did not occur.

In spite of this result, there was concern that, if a Monel firebreak section is used within a well bore, radiation from carbon steel burning underneath the Monel or heat transfer through conduction might preheat the Monel before the arrival of the flame front. This preheating could lower the Monel's threshold to the point where a fire might traverse through the firebreak section. Furthermore, burning carbon steel might be superior to thermite as an ignition source. To allay these concerns, several composite test specimens with a welded carbon steel-to-Monel interface were prepared and similarly tested in commercially pure oxygen. As shown in Fig. 9, complete destruction of the

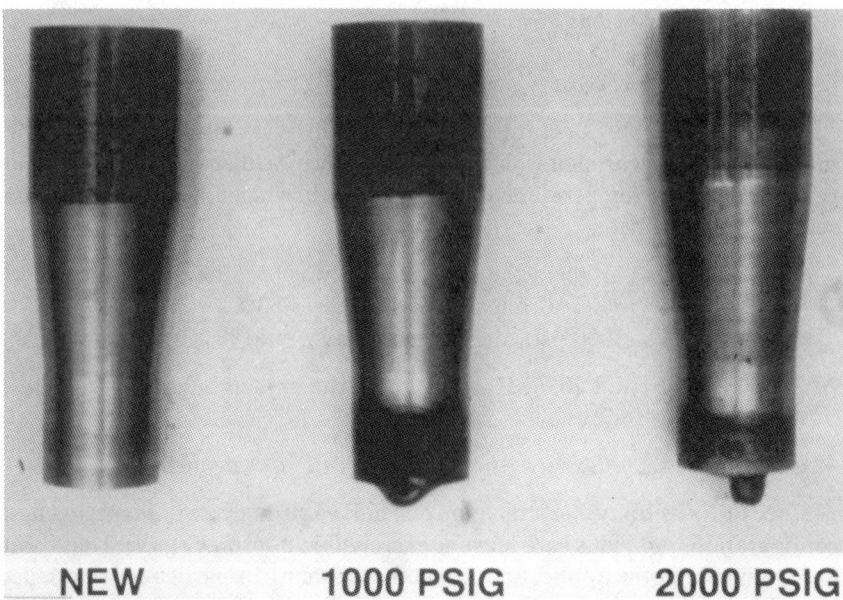

FIG. 8—*Monel combustion results in pure oxygen.*

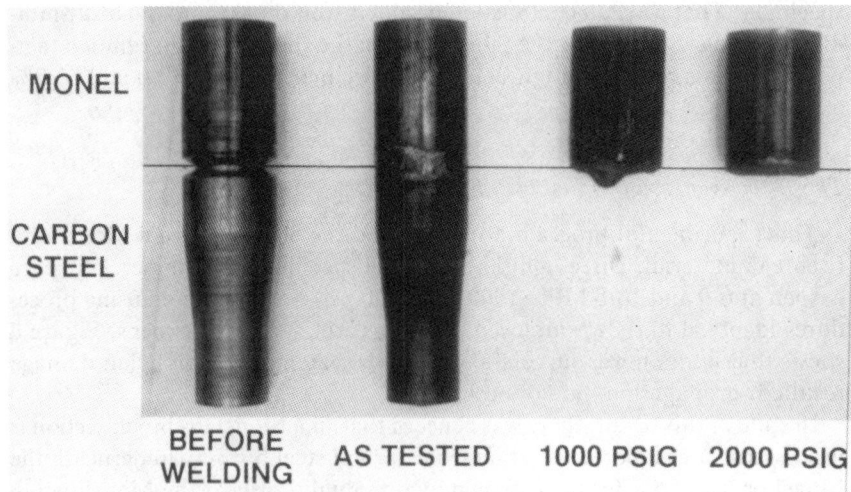

FIG. 9—*Composite specimen combustion results in pure oxygen.*

carbon steel occurred, but the combustion zone did not propagate into the Monel region. The test demonstrated the efficacy of Monel as a firebreak material.

Results and Conclusions

Oxygen Index Was Measured as a Function of Temperature and Pressure

For C-1018 carbon steel tube (25.4 mm [1 in.] outside diameter and 4.8 mm [³/₁₆ in.] wall) through which gas flowed downward at 0.30 m/s (1 ft/s) at roughly ambient temperature (300 K), the following upward propagation results were obtained:

- At 1.03 MPa (150 psi), the oxygen index was 81% oxygen in nitrogen.
- At 6.9 MPa (1000 psi), the oxygen index was 53%.
- At 20.7 MPa (3000 psi), the oxygen index was 51%.

At 355 K (180°F) and 20.7 MPa (3000 psi), the oxygen index was 47%.

Values of Oxygen Index Are Always Configuration Dependent

Since the possibility of combustion depends so strongly on the experimental configuration, judicious care must be exercised in interpreting these data and evaluating them for application to another system. In particular, one must consider the effects of significant changes in dimension, whether combustion is most likely to be up or down, what potential ignition sources there might be, and whether combustion of contaminants could generate enough preheating to significantly reduce the oxygen index of the material of interest. Also, note that the oxygen index is a lower limit for combustion to occur, not an upper limit to prevent combustion.

Gas Velocity Did Not Affect Oxygen Index

In tests near, but below, the oxygen index at 6.9 MPa, ignition at gas velocities one-half and three times the standard velocity of 0.3 m/s, damaged the specimens about equally but did not cause flame propagation. This suggested that the oxygen index was not significantly reduced at these gas velocities.

Burning Buna N Can Ignite Carbon Steel

Burning Buna N rubber in amounts at least as small as 0.95 g can ignite C-1018 alloy carbon steel at room temperature in oxygen concentrations at least as low as 53.2% at 6.9 MPa pressure.

A Complex Chain of Events Might be Required to Ignite Carbon Steel

Carbon steel exhibits complex ignition properties. A roughly 23-kJ (5500-cal) thermite ignition pill was capable of reliably igniting carbon steel specimens with, at most, three misfires in 16 tests conducted at concentrations equal to or above the ultimately defined oxygen indexes. However, a kindling chain mechanism was required to achieve this reliability. In fact, until the specimen configuration was finalized, these same thermite ignition pills did not reliably ignite any specimens at concentrations above the threshold, or even in pure oxygen. Furthermore, when ignited Buna N or crude oil of comparable or even greater heat release was substituted for the thermite pill in the final specimen configuration, the reliability was lost. Subsequent modifications to the configuration were necessary before successful ignition by Buna N was achieved.

Monel 400 Was an Effective Firebreak

At room temperature at pressures to 13.8 MPa (2000 psi), Monel 400 did not burn in commercially pure oxygen when ignited by a thermite pill. Futhermore, with a composite specimen of steel and Monel at the same test conditions, fire did not propagate into the Monel, even though the steel was completely consumed.

References

[1] Hvizdos, L. J., Howard J. V., and Roberts, G. W., "Enhanced Oil Recovery Through Oxygen-Enriched In Situ Combustion: Test Results from Forest Hill Field in Texas," *Journal of Petroleum Technology*, Vol. 35, No. 7, June 1983, pp. 1061–1070.

[2] Benning, M. A., "Measurement of Oxygen Index at Elevated Pressures," *Flammability and Sensitivity of Materials in Oxygen-Enriched Atmospheres, STP 812*, B. L. Werley, Ed., American Society for Testing and Materials, Philadelphia, 1983, pp. 68-83.

[3] Hendrix, J. E., Drake, G. L., Jr., and Reeves, W. A., "Effects of Temperature on Oxygen Index Values," *Textile Research Journal*, Vol. 41, No. 4, April 1971, p. 360.

[4] Wells, A. A., "The Iron-Oxygen Combustion Process—A Study Related to Oxygen Cutting," *British Welding Journal*, Sept. 1955, p. 392.

[5] Harrison, P. L. and Yoffe, A. D., "The Burning of Metals," *Proceedings of the Royal Society of London, Series A*, Vol. 261, 1961, p. 357.

[6] Friant, C. W., "Determination of the Burning Rate of Small Diameter Aluminum Wire, M.S. thesis, Virginia Polytechnic Institute, Blacksburg, VA, June 1964.

[7] Dean, L. E. and Thompson, W. R., "Ignition Characteristics of Metals and Alloys," *ARS Journal*, July 1961, p. 917.

[8] McKinley, C., "Experimental Ignition and Combustion of Metals," Oxygen Compressors and Pumps Symposium, Compressed Gas Association, 9, 10, and 11, Nov. 1971, p. 27.

[9] Kirschfeld, L., "Combustibility of Steel and Cast Iron in Oxygen at Pressures of up to 150 Atms," *Archiv f.d. Eisenhuttenw*, Vol. 39, No. 7, 1968, p. 535.

[10] Kirschfeld, L., "Combustibility of Metals in Oxygen," *Angewandte Chemie*, Vol. 71, No. 21, 1959, p. 663.

[11] Kirschfeld, L., "Apparatus for Combustion Tests on Metals Under Oxygen Pressures Up to 200 Atm, and the "Combustibility of Iron Wire in High Pressure Oxygen," *Archiv f.d. Eisenhuttenw*, Vol. 36, No. 11, Nov. 1965, p. 823.

[12] Lowrie, R., "Heat of Combustion and Oxygen Compatibility," *Flammability and Sensitivity of Materials in Oxygen-Enriched Atmospheres, STP 812*, B. L. Werley, Ed., American Society for Testing and Materials, Philadelphia, 1983, pp. 84-96.

John W. Grubb[1]

Case Study of Royal Australian Air Force P3B Orion Aircraft Ground Oxygen Fire Incident

REFERENCE: Grubb, J. W., **"Case Study of Royal Australian Air Force P3B Orion Aircraft Ground Oxygen Fire Incident,"** *Flammability and Sensitivity of Materials in Oxygen-Enriched Atmospheres: Second Volume, ASTM STP 910*, M. A. Benning, Ed., American Society for Testing and Materials, Philadelphia, 1986, pp. 171–179.

ABSTRACT: In Jan. 1984, the Royal Australian Air Force (RAAF) experienced a ground fire incident that destroyed a six million dollar P3B Orion Aircraft (A9-300). The incident occurred during removal of an onboard oxygen cylinder, one of three that supply the Orion flight crew. Examination of the aircraft's oxygen system revealed that the fire had initiated in an oxygen manifold check valve (MCV) assembly. The primary cause of the incident was a leaking poppet valve, which allowed oxygen stored at 12 MPa (1800 psi) to escape to atmosphere. Deterioration of the silicone rubber seal and galvanic corrosion are believed responsible for the valve failure. Contributory causes to the fire were system contamination and failure to bleed the oxygen system before cylinder disconnection. A thermite reaction involving the aluminum MCV assembly, metal particles, and metal oxide particles was the most likely cause of ignition. RAAF investigations revealed similar contamination throughout the entire fleet of P3B/C Orion aircraft; iron oxide dust was the major contaminant found. Oxygen system filters were also examined, and trials aimed at improving particulate matter filtration are currently being conducted. Extensive cleaning of both aircraft oxygen systems and ground support equipment was also undertaken.

RAAF findings, supported more recently by the U.S. Navy investigations, indicate the need to consider using materials other than silicone rubber in oxygen systems. Further research into the ignition of aluminum and other candidate materials through metal particle impingement in the presence of metal oxides in a high-pressure oxygen environment is also required.

KEY WORDS: ground fire, Orion aircraft, oxygen, silicone, galvanic corrosion, contamination, thermite reaction, metal oxide, filtration

In Jan. 1984, the Royal Australian Air Force (RAAF) experienced a ground oxygen fire incident that destroyed a six million dollar P3B Orion Aircraft

[1]Squadron leader, Staff officer aircraft equipment engineering, Royal Australian Air Force, Embassy of Australia, 1601 Massachusetts Ave. NW, Washington, DC 20036.

(A9-300). This paper presents details of the incident, ensuing RAAF investigations and Court of Inquiry Findings.

Description of P3B Orion Oxygen System

A schematic diagram of the Orion oxygen system is shown in Fig. 1. Oxygen stored in three high-pressure cylinders at 12 MPa (1800 psi) supplies three regulators, one for each flight crew member. A high-pressure oxygen manifold assembly equipped with internal check valves receives flow from the cylinders and directs it via a common line to a 0.43-MPa (65-psi) pressure reducer. The manifold assembly also connects to a filler line allowing the three cylinders to be charged simultaneously from an external supply.

Oxygen Fire Incident: Court of Inquiry Findings

The fire occurred when the center oxygen cylinder under the pilot's seat was being removed to gain access to the aircraft's forward radar fixed wave guide. The flexible hose end fitting connected to the cylinder had been loosened beforehand to bleed the oxygen line. Loosening this end fitting also allowed a check valve fitted to the cylinder to close. Contrary to technical maintenance

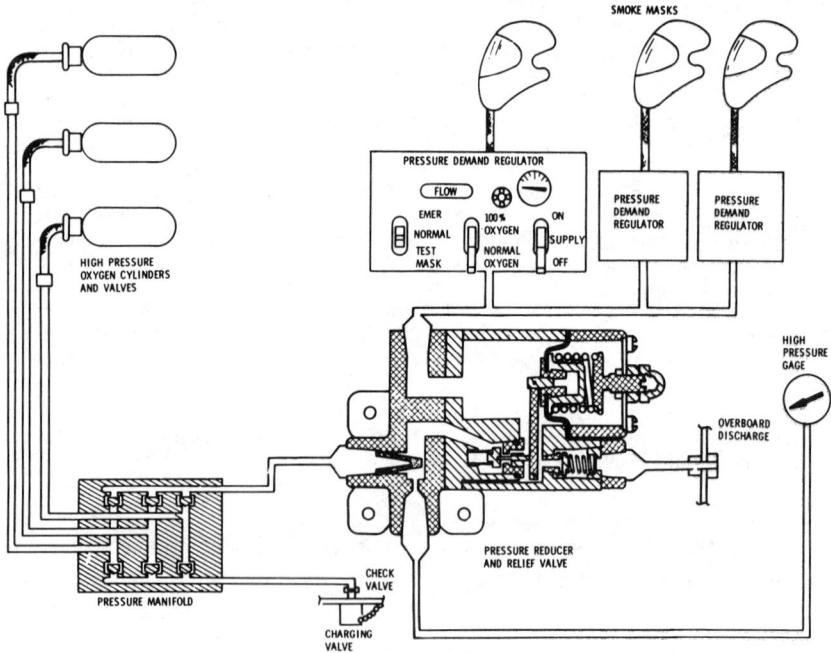

FIG. 1—P3 *Orion aircraft oxygen system schematic.*

instructions, the oxygen system had not been bled from 12 to 3.33 MPa (1800 to 500 psi), nor had the emergency exit on the port side of the cockpit been opened.

Primary and Contributory Causes of Incident

Examination of the aircraft's oxygen system revealed that the fire had initiated in the high-pressure oxygen manifold check valve (MCV) assembly. The primary cause of the incident was determined to be a leaking poppet valve in the center outlet chamber of the pressure manifold. The leaking valve allowed a reverse flow of high-pressure oxygen to atmosphere when the flexible hose end fitting was loosened. Contributory causes were determined to be the presence of contaminants and failure to bleed the oxygen system to 3.33 MPa (500 psi).

RAAF findings [1] suggest that the cause of ignition for the fire in A9-300 was a thermite reaction involving aluminum/iron with iron or copper oxides or both. Energy for the initiation of the reaction probably came from metal particle impact associated with high gas velocities present in the oxygen system caused by the leaking poppet valve.

Figure 2 shows the cross section of an MCV assembly removed from RAAF P3C aircraft A9-756. The photograph shows evidence of an incipient fire in the inlet chamber, just below the inlet poppet valve seat. Iron oxide dust, though not easily distinguished, is present on the chamber walls. As in A9-300, a thermite reaction is believed to be the cause of ignition. Fortunately, the ignition in A9-756 did not propagate to become a destructive event.

Poppet Valve Failure

The type of MCV assembly poppet valve used in A9-300 is shown in Fig. 3. The poppet valve comprises a silicone rubber seat[2] bonded to an aluminum body and secured by an anodized aluminum washer and leaded brass retaining screw. Figure 4 shows the poppet valve removed from A9-300. Intergranular corrosion is present over its entire length. The head of the seal retaining screw is missing; failure caused by galvanic corrosion between the aluminum washer and brass retaining screw is believed to have occurred before the fire.

Oxygen System Contamination

A thorough examination of the aircraft's oxygen system revealed the following contaminants:

---

[2]Silicone Rubber Spec ZZ-R-765, Class 1A-50 manufactured by Dow Corning PN 6514, composition (parts per 100) 89 parts siloxane, 10 parts fused silica filler, 1 part iron oxide thermo stability improver.

FIG. 2—*Cross section of MCV assembly: RAAF P3C aircraft A9-756.*

(1) coarse particles (10 to 500 μm) comprising copper based alloys, silicas, nonmetallics, free iron, and steel and aluminum;

(2) fine particles (<1 μm) of iron oxide; and

(3) corrosion products, including brass dust, zinc oxide, copper, and copper oxides from corrosion of the poppet valve retaining screw.

Post Oxygen Fire Investigations/Findings

Having established the primary and contributory causes of the incident, the RAAF directed further investigations to the following areas:

(1) source and extent of contamination in Orion aircraft oxygen systems and ground support equipment,

(2) effectiveness of Orion oxygen system filters, and

(3) extent to which poppet valves utilizing brass retaining screws were fitted to Orion oxygen MCV assemblies.

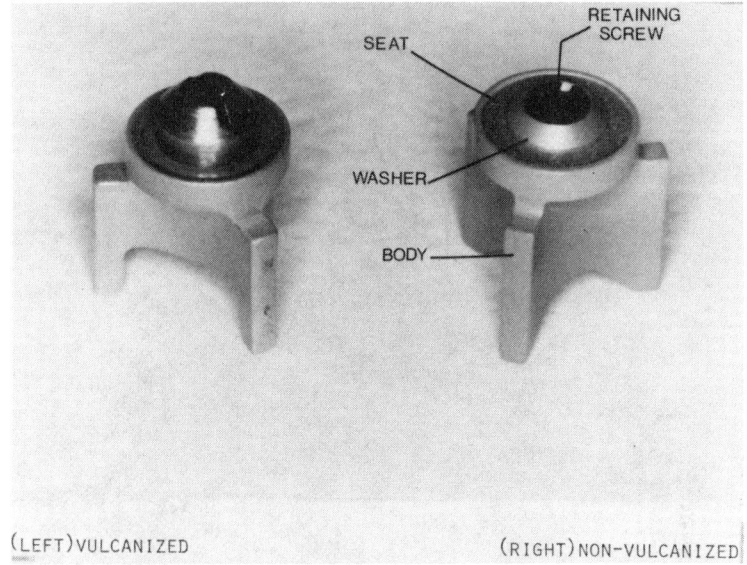

(LEFT) VULCANIZED   (RIGHT) NON-VULCANIZED

FIG. 3—*Types of poppet valve fitted to RAAF P3B/C oxygen systems.* Left: *vulcanized.*
Right: *nonvulcanized.*

Contamination

Examination of the entire RAAF fleet of P3B/C aircraft revealed oxygen system contamination similar to that found in A9-300. In some cases iron oxide dust was found beyond the pressure reducing valve. Coarse contaminants were present in both recharging and high-pressure oxygen lines. Spot checks of other RAAF aircraft (for example, DHC4 CARIBOU and MIRAGE) revealed similar oxygen system contamination.

During the course of investigations, the RAAF also found that the 0 ring in the oxygen cylinder valve-coupling assembly sustained damage during disconnection and reconnection. Approximately 80% of aircraft oxygen cylinders had worn 0 rings and neoprene particles present in the neck of the cylinder. Another contributing factor to system contamination was the mismatch of brass and steel fittings on P3B/C oxygen charging connections and charging equipment.

Examination of ground support equipment revealed iron oxide contamination in ground transportation oxygen cylinders and oxygen charging trolleys. Subsequent investigations isolated the source of the iron oxide dust to poorly maintained transportation cylinders.

FIG. 4—*Poppet valve removed from Orion aircraft A9-300.*

P3B/C Orion Oxygen Systems Filters

Figure 5 shows the charging valve filter element found in A9-300. The nominal grade of this filter is 125 μm; the filter spec requires 65 μm. Figure 5 also shows the charging valve recess where the filter element base contacts the valve body. The dark spots (corrosion) indicate where the filter element made contact with the valve Body. Gaps between the element and the valve body were found to be several times larger than 125 μm. Inspection of the P3B/C fleet revealed fitment of both 125- and 65-μm elements. Both coarse and fine filters were examined to ascertain the nature and particle size of the debris trapped and passed by the filter. The results shown in Table 1 do not favor either filter. RAAF believes the majority of contaminants found in the P3B/C Orion fleet oxygen systems were introduced externally through the charging valve filter. The inlet filter to the pressure reducer on P3B/C Orion aircraft was also examined. RAAF investigations discovered that the filter retaining clip in some reducers did not adequately seat the filter element, hence negating filter efficiency.

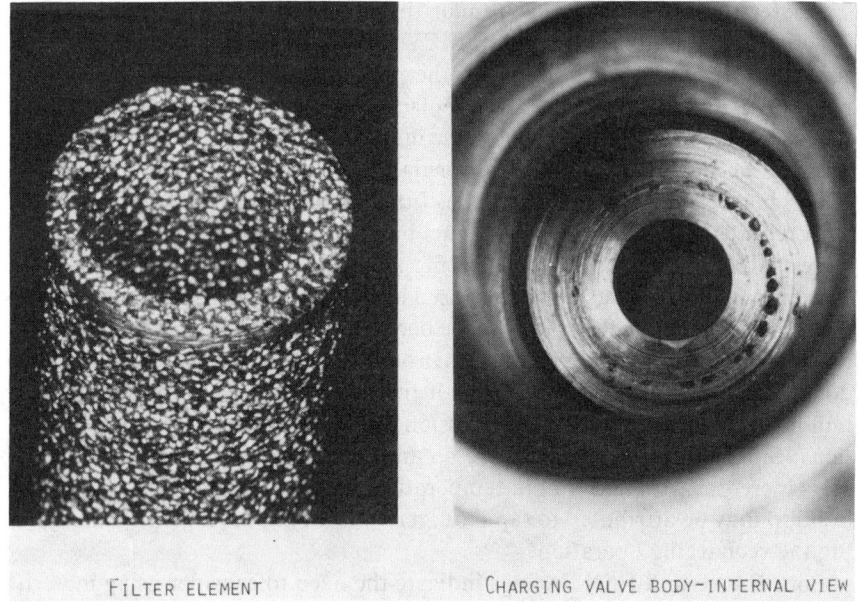

FILTER ELEMENT   CHARGING VALVE BODY-INTERNAL VIEW

FIG. 5—*Filter element and charging valve body removed from Orion aircraft A9-3000.*

TABLE 1—*Results of ARL examination of fine and coarse filters*
(oxygen filler valve assembly).

Filter Size, μm	Particulate Matter Removed From Filter, μm		Particulate Matter Passed Through Filter, μm	
	Maximum	Minimum	Maximum	Minimum
125	400	50	100	1.0
65	200	50	100	1.0

[a]Particulate matter removed consisted of brass, silica, aluminum, and non-metallics.

Poppet Valve: P3B/C MCV Assemblies

Investigations revealed two variants of MCV poppet valve fitted to P3B aircraft oxygen systems. Poppet valves comprising both leaded and unleaded brass retaining screws were found. Both variants exhibit a number of defects:

1. The silicone rubber seal tends to separate from the poppet body at the edge and crack.

2. Galvanic corrosion occurs between the aluminum washer and brass retaining screw.

3. The sharp edge of the washer can cut into the silicone rubber seal.

A later type of poppet valve (under the same part number as the earlier variants) was found in P3C aircraft MCV assemblies (Fig. 3). This type has a silicone rubber seat vulcanized into the poppet valve body, with the aluminum washer and brass screw being replaced by a large headed stainless steel screw. The new style seal uses the same material as the old seal. Based upon a recently published Lockheed California Company Report [2], U.S. Navy (USN) inspections of MCV assemblies fitted to both P3B and P3C Orion aircraft revealed examples of embrittlement and erosion on the new style poppet valve seal. USN conclusions, as cited in the Lockheed Report, were that deterioration of the seals was possibly caused by two phenomena. First, embrittlement of the silicone rubber took place because of cross-linking of the silicone polymer chain in the oxygen rich environment. Second, regression (breakdown of the silicone polymer chain) of the seals possibly occurred with heat buildup caused by adiabatic compression of oxygen during recharging operations. Since these phenomena have a cumulative effect, the deterioration of seals increases with age. Rapid temperature rise, and hence increase in heat buildup may be attributed to rapid oxygen flow rate under high pressure during the recharging operation.

Both RAAF and USN findings indicate the need to consider using materials other than silicone rubber in an oxygen environment. Inert synthetic materials such as Kel-F or Viton (fluorocarbon and elastomer) have been recommended by USN [2].

Use of Aluminum in High-Pressure Oxygen Environments

Although the RAAF did not address it in the investigation, the extensive use of aluminum in high-pressure oxygen systems is an issue worthy of further attention. The choice of materials for these applications is a compromise between many factors, including safety, weight, strength, corrosion resistance, and cost. This incident questions the wisdom of using aluminum in this type of environment.

Further investigations into the ignition of aluminum and other candidate materials through metal particles impingement in the presence of metal oxides (for example, rust, copper, or chromium oxides) is considered warranted; particularly so, since assured elimination of contaminants is probably unattainable. Out of this might come a range of practical measures for reducing the risk of ignition, at the very least, guidelines to future designers as regards material selection.

Actions Stemming from RAAF Investigations

Aircraft Oxygen Systems

In view of the extensive iron oxide dust contamination discovered throughout the RAAF P3B/C Orion fleet, all oxygen systems were disabled and portable oxygen sets installed. Recommissioning occurred only after a thorough cleaning and purging of the oxygen system had been undertaken. Use of the old type poppet valve has been discontinued, and new style poppet valves have been installed. All MCV assembly poppet valves will now be replaced after each one installed year. Charging valve filters (65 μm) are now fitted throughout the P3B/C fleet and inspected, cleaned, or replaced coincident with servicing of the MCV assembly poppet valves. In addition, Orion aircraft oxygen systems are now bled to 0.66 MPa (100 psi) before maintenance.

GSE and Maintenance Installations

In order to prevent recontamination of aircraft oxygen systems an inspection and cleaning program of all RAAF owned transportation cylinders has been initiated. Cylinders are steam cleaned with a high alkaline detergent, hot rinsed, and dried with dry nitrogen. In addition, all aircraft charging equipment was thoroughly inspected and cleared for use with recertified cylinders. Protective treatment of internal cylinder surfaces with an inhibitor is also under consideration. Currently, the RAAF is evaluating 5-μm filters installed in oxygen charging equipment.

Conclusions

The RAAF lost a six million dollar aircraft through the presence of a defective component. Deterioration of a silicone rubber seal and galvanic corrosion are believed responsible for component failure. Contributing factors associated with the incident were system contamination and the application of incorrect maintenance procedures. A thermite reaction involving the aluminum MCV assembly, metal particles, and metal oxide particles was the most likely cause of ignition. Further research into the ignition of aluminum and other candidate materials through metal particle impingement in the presence of metal oxides in a high-pressure oxygen environment is required.

RAAF findings, supported more recently by USN investigations, indicate the need to consider using materials other than silicone rubber in oxygen systems. The incident highlights the need for continued emphasis on high-maintenance standards associated with the carriage and handling of oxygen.

References

[1] Barter, S. A. and Hillen, L. W., "Report to Court of Inquiry into the Cause of a Fire in Orion Aircraft A9-300," Aeronautical Research Laboratory, Melbourne, Australia, Feb. 1985.
[2] Seino, L., "P-3 Flight Station Oxygen System," white paper, CALAC Engineering, Lockheed California Company, 19 Dec. 1984.

Dennis W. Schroll[1]

Cleaning Methods and Procedures for Military Oxygen Equipment: Investigation Results

REFERENCE: Schroll, D. W., **"Cleaning Methods and Procedures for Military Oxygen Equipment: Investigation Results,"** *Flammability and Sensitivity of Materials in Oxygen-Enriched Atmospheres: Second Symposium, ASTM STP 910,* M. A. Benning, Ed., American Society for Testing and Materials, Philadelphia, 1986, pp. 180–203.

ABSTRACT: Over the past several years a number of briefings have been given at the Oxygen Standardization Coordinating Group (OSCG) concerning cleaning methods and procedures for military oxygen equipment. Of particular concern has been the diversity of methods, cleaning compounds, and cleanliness criteria that exist in the military and industry in the manufacture, shipment, aircraft installation, and maintenance of military oxygen equipment. This will be discussed in some depth in this paper. The OSCG is an Army, Navy, and Air Force group that meets twice a year to discuss and coordinate changes to specification governing procurement of military oxygen equipment and related design and installation practices. Industry is also invited to attend and comment.

KEY WORDS: oxygen components, oxygen compatibility, oxygen cleanliness, cleanliness standards, oxygen purity, aircraft fire, flammability, contamination, oxygen compatibility, cleaning laboratories

At the 40th Oxygen Standardization Coordinating Group (OSCG) meeting, held June 1983, considerable discussion centered around Military Standard Cleaning Methods and Procedures for Breathing Oxygen Equipment (MIL-STD-1359A) and proper oxygen cleaning procedures. The standardization office at the Federal Center, Battle Creek, MI, sent a request for information on 11 July 1983 to 58 industrial and military organizations. This survey requested information that included a description of their cleaning procedures, what document is used to support the procedures, and any "lessons learned" that may be of benefit to an evaluation. Four major aircraft companies, four major oxygen equipment manufacturers, one oxygen equip-

[1] Senior life support engineer, U.S. Air Force Aeronautical Systems Division, ASD/ENECE, Wright-Patterson AFB, OH 45433.

ment test agency, and two military organizations dealing with oxygen equipment all provided an outstanding response. The standardization office requested that these comments and cleaning specifications be sent to the author of this paper for evaluation and a final disposition. A briefing was given at the 41st OSCG meeting to summarize tentative conclusions and recommendations on this issue. Also, many other briefings on this issue have since been given by other individuals [1–4].

Scope

Much of this activity has centered around whether there is a need to make changes in military cleanliness requirements. Changes have been proposed that would standardize many aspects of cleaning. It has been proposed that a new military document, such as a handbook or standard, be established. All other requirements used would emanate from this document. Considerable investigations have been conducted that deal with cleaning requirements of oxygen equipment. Recently, the oxygen equipment maintenance organization at Tinker Air Force Base, near Oklahoma City, has had numerous meetings with military representatives of each aircraft command and systems managers of the aircraft equipment. Revised aircraft oxygen systems purging procedures have been determined. Further work is also underway to develop new and improved aircraft purge units.

Additionally, investigations by oxygen engineering experts at Wright-Patterson Air Force Base have revealed the need to revise purging equipment and requirements for liquid oxygen ground storage tanks and transport carts. Some time ago NASA conducted a similar oxygen systems cleaning survey [5]. This survey was primarily concerned with cleanliness of rocket systems liquid oxygen storage units and functional components of space equipment.

These investigations, however, represent only a fraction of the total oxygen equipment cleaning issue. This includes cleaning aircraft oxygen systems, aircraft oxygen ground support equipment, maintenance laboratory cleaning requirements, military overhaul facilities of precision oxygen equipment (the so-called Precision Measurement Equipment Laboratories [PMEL's]), design and development cleaning requirements of oxygen equipment, and the cleanliness requirements at the industry production and assembly lines.

Cleaning, inspection, and packaging requirements are established from the government to industry by military specification and standards. That information which governs the proper operation of military facilities is usually governed by military Technical Orders (or T.O.'s) [6].

This paper will focus on the first and last concern. That is, cleaning aircraft oxygen systems, and military guidance for industry oxygen equipment cleaning, inspection, and packaging requirements. Should a military document be developed that standardizes cleaning requirements, it should address the following areas:

(1) definition of contaminant levels allowable in oxygen equipment,

(2) general requirements for cleaning oxygen equipment components,

(3) special cleaning instructions for different types of oxygen equipment,

(4) inspection methods for verification of cleanliness and cleanliness levels applicable,

(5) package requirements and methods to prevent recontamination, and

(6) special methods for cleaning and verification at the aircraft.

The following text represents a composite of information from this survey. That information provided in the text is considered the desirable techniques in cleaning oxygen equipment for U.S. Air Force purposes. Additionally, a summary of all cleaning methods used (acceptable and unacceptable) are given in Tables 2 and 3 for background information and comparison purposes.

Military Guidance for Cleanliness Criteria in Industry

Contaminant Levels Allowed

The contaminants levels allowed in the oxygen equipment should be defined. Some of these contaminants are hydrocarbons, particulate solids, moisture, and cleaning solvents. The contaminant level allowed will vary with the type of oxygen equipment and in which state the oxygen is used. In other words, the contamination problem will be different for liquid oxygen (LOX) versus gaseous oxygen (GOX) equipment. For ambient temperature ranges of operation contaminant levels allowed will vary for four pressure ranges of operation in GOX equipment and delivery components of LOX systems:

(1) low pressure range (0 to 345 kPa) (0 to 50 psig),

(2) working pressure range (345 to 3450 kPa) (50 to 500 psig),

(3) high-pressure range (3450 to 20 680 kPa) (500 to 3000 psig), and

(4) extreme high-pressure range (20 680 to 103 420 kPa) (3000 to 15 000 psig).

In general, the higher the pressure and temperature the cleaner the equipment must be to prevent fire and explosions. The allowable contaminant level is a function of the specific chemical composition of the contaminant. Much work remains to be accomplished in this area to delineate allowable levels of contamination for specific substances. Contaminants must also be minimized to eliminate odors in breathing oxygen gas. Also, moisture will freeze in delivery components and preclude effective operation.

Environmental Considerations in Cleanliness

An area of concern in cleaning oxygen equipment is the environment or work area cleaning room. It would not be possible to clean the equipment in an environment that has airborne contaminants that adhere to the surface of the oxygen equipment immediately after it has been cleaned and retrieved from a solvent solution. To ensure that oxygen equipment is properly cleaned and remains clean until properly packaged, it is essential to maintain a controlled environment to prevent recontamination during cleaning, drying, assembly, testing, and packaging. For high levels of environmental cleanliness, the methods and techniques described in Clean Room and Work Station Requirements, Controlled Environment (FED-STD-209) may be used to achieve the desired level of environmental control for a work area. The work room cleanliness practices that should be followed to achieve a high degree of cleanliness are as follows [Ref 6 and MIL-STD-1359A)]

1. Environmental control: The work area should be maintained at a temperature of about 22°C (72 ± 5°F) with a relative humidity of about 35 ± 15%.

2. Room air pressure levels: All rooms used for cleaning, drying, assembly, testing, inspection, and packaging of oxygen equipment should be designed and outfitted to maintain a pressure above that of the outside or surrounding areas. The intent is to assure that all air leakage will be outward of the clean room(s). Air conditioning equipment with appropriate air filters should satisfy this requirement.

3. Air flow and ventilation: A constant air flow of filtered air is desired throughout the clean room(s). Recirculating air and outside vent air must be filtered through a high efficiency particulate filter. When cleaning solvent fumes are not present, a recirculating air flow feature minimizes contamination in that air filtration requirements are usually reduced. When air ventilation is required for removal of toxic and noxious fumes an outside air vent feature should be incorporated. The design of the air flow system should provide a laminar flow of air across the work area where possible. Stagnant air locations should be eliminated from the work areas.

4. Maintenance equipment and tools: Work benches, tools, and any cleaning equipment should be kept free of grease, oil, or any suspected combustible materials. Tools used for maintenance of oxygen equipment should not be used for any other purpose. Any paper used in the clean rooms should be nonshedding and lint-free to preclude the adhering of an organic substance to clean oxygen equipment. Pencils that contain graphite induce corrosion on metals. Therefore only pens and pencils without graphite should be used in the clean rooms.

5. Personnel cleanliness: Personnel who work in the clean rooms are expected to maintain themselves and their clothing in a condition that prevents transferring contaminants to any oxygen equipment. It is generally desirable that personnel working in clean rooms wear special clothing that is lint-free. A white garment will show contamination easily, and its use is encouraged. Smoking, eating, or any activity that permits contamination of the clean rooms should not be allowed.

6. Work room cleaning requirements: The clean room area must be maintained in a clean and dust-free condition. A vacuum cleaner that does not vent airborne particles should be used to remove dust and dirt. Damp mopping should be used for the local removal of adhering dust and dirt. All parts in the cleaning process should be removed from work benches or covered with lint-free covering at the last work shift each day. All tools should be put away clean. Work benches and test equipment should be cleaned with a lint-free cloth at the start of each work shift. It should be noted that many cleaning environments used to clean oxygen equipment have been established in the military and industry that do not provide the level of cleanliness in this type of clean room. This may be acceptable considering the degree of cleanliness desired.

The Cleaning Process

Cleaning methods, procedures, and standards must be established that are safe, nontoxic, and not detrimental to the oxygen equipment being cleaned. Keeping in mind that the function of a cleaning agent is to assist in the removal of organic and inorganic material from the oxygen equipment, certain guidelines have been developed to consider all factors. A cleaning agent is good when

(1) it is successful in removing contaminants,
(2) it does not leave residues or solids,
(3) it is relatively nonflammable,
(4) it is relatively low in toxicity,
(5) it will not react chemically or attack parts being cleaned, and
(6) it will not soften or distort any materials being cleaned.

The most commonly used solvent for cleaning metals and Teflon® trichlorotrifluoroethane, MIL-C-81302, Type I or II, and soaps and detergents are used for other nonmetals such as rubber, silicone, and gaskets. Methyl chloroform, O-T-620, and 1,1,1 trichloroethane, MIL-T-81533A, are also frequently used to clean metals and Teflon®.

Aids to assist cleaning would be some means of agitating the cleaning agent in contact with the surface being cleaned or scrubbing the surface with brushes or both. Equipment that has been constructed with complex openings and surfaces, such as oxygen regulators, may require special attention to ensure that contaminants have not been overlooked during the cleaning process. It should also be kept in mind that the cleaning solvent should only be used for that application for which it is intended (that is, vapor degreasing, solvent, and ultrasonics). Steps that should be included in the cleaning process are as follows:

1. Precleaning or gross cleaning.
2. Removal of oxide layers from metals, if needed.

3. Cleaning and contaminant removal by a solvent flush or total immersion in the proper cleaning solvent and agitation, if needed.

4. Cleanliness inspection that may require special techniques and equipment.

5. The complete removal and drying of the cleaning agent from the part. This may include water flushing and purging with dry, clean nitrogen or air gas. The final step in purging an installed breathing system should always include purging with dry, clean oxygen gas. Safe pressure and temperature limits should not be exceeded.

6. Post drying inspection of the cleaned equipment. Also check for solvent removal with a halogen detector or by an equivalent method.

7. Next the parts and components need to be packaged to avoid recontamination. The package material should be cleaned to the same level of cleanliness as the part. This includes the use of plastic bags that are sealed about the components and the use of plugs in part holes.

Special Cleaning Instructions for Different Types of Oxygen Equipment

Oxygen system components, installation plumbing lines, and valves can be more easily cleaned when broken down to components and parts. These different parts may contain plastic or elastomeric parts as well as gaskets, diaphragms, and O-ring seals that are subject to deterioration and damage when exposed to some cleaning agents. Additionally, resulting odors may be nearly impossible to eliminate as the porosity of the part may off-gas solvent odors for many months even after flushing. Breaking the oxygen system down to components and parts allows each part to be cleaned in the way most effective for its constituent material. Additionally, cleaning before assembly of parts or components assists greatly in achieving a clean assembly.

Another reason disassembly assists in cleaning concerns the accumulation of oils, dust, metal chip, metal oxides, and so forth that cannot be effectively removed by the hot gas purging method normally used to clean oxygen systems. Liquid flushing may also be used to attempt to remove contaminants, but particles will become trapped in a narrow orifice or low gravity trap. The trapped contaminants may preclude the proper operation of the oxygen system and may even lead to a fire or explosion. Such contaminants can only be removed by the complete disassembly of the affected components. With proper cleaning procedures, solvents, and the proper packaging of the components, the oxygen system may be reinstalled such that it is effectively clean.

Cleaning materials and processes selected should be compatible with the components and parts to be cleaned. Solutions and solvents used for cleaning and gases used for purging and drying should be of such purity such that contaminants and undesirable residues will not be left after the final drying process. All cleaning operations are complex operations that require the direct supervision of a trained individual with experience involving the components being cleaned, the cleaning materials being used, the cleaning equipment in use, and the safety precautions associated with the entire cleaning operation. Individuals involved in the cleaning operation should use protective equipment for the face, eyes, hands, and other parts of the body subject to hazards. Of special concern is that some cleaning solvents will give off toxic fumes. This means proper ventilation is essential to prevent poisoning people and combustion of the air solvent gas mixture.

Of primary concern in cleaning aircraft oxygen systems and ground support equipment for the aircraft are the following components: liquid oxygen converter, oxygen mask, personnel breathing hoses, breathing oxygen regulators and valves, oxygen transfer hose and lines, oxygen cylinders and associated valves, pressure gage, liquid oxygen storage and transfer tank, and parts of all of these components after disassembly. Table 1 gives a cleaning method and agent for each type of equipment to be cleaned as well as its current military practice. Table 2 gives alternative materials and cleaning agents used in the military and industry.

Inspection Methods for Verification of Cleanliness and Cleanliness Levels Applicable

The techniques and solvents involved in cleaning oxygen system components vary with the properties of the material (for example, metallic versus non-metallic) being cleaned. After cleaning, however, an inspection method is necessary to determine that cleanliness has been achieved. When the desired level of cleanliness has been achieved, then a technique for removing all cleaning agents used must be used. Water soluble cleaning agents may be removed by flushing or agitation with ample quantities of distilled or de-ionized water. This ensures that deposits that may be in solution in common tap water will not leave impurities and mineral deposits on the cleaned oxygen component. The equipment should then be dried. There are various methods for purging and drying, but in all cases only gases that are clean at least to the level of the equipment cleaned should be used. Methods of reducing the time required for drying have been developed and consist of oven drying and purging with heated gases. In most military oxygen equipment cleaning operations gases used in purging and drying are oil-free air per BB-A-1034, nitrogen per BB-N-411, Type I, Class I, Grade B, and oxygen per MIL-O-27210. Purging and oven drying temperatures are specified not to exceed 121°C (250°F) in general and even less with some temperature sensitive materials. After drying, a post cleanliness inspection is then necessary to ensure that all solvent and solvent removal solution have been removed. A common technique to check for the complete removal of solvent residues is with a halogen detector. This method will detect any cleaning solvent of the halide group left as residue.

The results of cleaning are not normally visible to the unaided eye, how-

TABLE 1—*Cleaning of the individual type of military oxygen equipment.*

Equipment MIL-STD-1359A	Cleaning Method	Cleaning Agent
Liquid oxygen converter	1. Purging method.	
	2. Purge with gas not to exceed 250°F at the inlet and a pressure of 30 ± 5 psig.	Use nitrogen gas, BB-N-411, Type I, Class I, Grade B or oxygen, MIL-O-27210, Type I
	3. Purge for at least 1 h and test outlet for odors.	
	4. If odors persist, purge longer. If odor still persists, flush converter with solvent.	
	5. Flushing method.	Flushing solvent, MIL-C-81302, Type I, Trichlorotrifluoroethane
	6. Flush solvent through the converter.	
	7. Test solvent for contaminants with halogin detector. Repeat flush if necessary.	
	8. Use purging steps 1 thru 4 to remove solvent.	
	9. After purging charge with oxygen gas at 30 psig.	Use oxygen gas MIL-O-27210, Type I
LOX storage containers	1. Warm inner tank above freezing temperature of the solvent with warm purging gas.	Air BB-A-1034 or nitrogen BB-N-411, clean and dry
	2. Heat degreasing solvent and circulate into the vessel.	MIL-C-81302, Type I or II trichlorotrifluoroethane
	3. Continue until inner tank reaches boiling temperature of solvent.	
	4. Purge tank again to remove all solvent.	Air, BB-A-1034 or nitrogen, BB-N-411, heated at inlet to 250 ± 25°F. Purge minimum 4 to 5 h until outlet temperature reaches 175°F.
Aircraft oxygen cylinders	1. Remove corrosion.	Use precleaning method
	2. Flush inside the cylinder with solvent spray.	MIL-C-81302 Type I trichlorotrifluoroethane
	3. Fill cylinder, plug end, and agitate solution. Clean out solution and repeat if necessary.	

TABLE 1—(Continued).

Equipment MIL-STD-1359A	Cleaning Method	Cleaning Agent
	4. Purge cylinder. 5. Dry cylinder.	Nitrogen, BB-N-411 at 250 ± 25°F for 1 h then fill with MIL-O-27210 oxygen at 30 ± 5 psig. Evaluate cylinder, heat in oven to 180°F for 1 h—fill to 30 ± 5 psig with oxygen.
Oxygen transfer hoses and lines (Teflon and Metallic) (NOTE: If nonmetallic material is unknown, use water soluble cleaning agent.)	1. Using cleaning solvent, flush and agitate with ultrasonics. 2. Circulate solvent through hose for at least 5 min. 3. Repeat procedure until contaminant level satisfactory. 4. When clean purge hose.	MIL-C-81302, Type I trichlorotrifluoroethane Use nitrogen gas, BB-N-411, or air, BB-A-1034 that is clean and dry.
Oxygen breathing mask inhalation and exhalation breathing valves	1. Separate mask, hose, valves and microphone components. 2. Wash mask and valves with a soap solution. 3. Lint-free cloth and soft bristle brush is used for cleaning. 4. Use water moistened cloth or sponge to remove soap.	. . . Use P-S-600 soap, 1 oz/gal of water or detergent, or MIL-D-16791, Type I, 1/4 to 1/2 oz/gal of water. Use distilled or demineralized water.

Component	Procedure	Cleaning agent
Breathing oxygen regulators and liquid oxygen converter parts	5. Dry with ambient air or low pressure gas stream.	Use only nitrogen gas, BB-N-411, or air, BB-A-1034.
	1. Disassemble into the smallest components.	...
	2. Clean in solvent by immersion with agitation or ultrasonics.	MIL-C-81302, Type I trichlorotrifluoroethane
	3. Inspect for contaminants and repeat if necessary.	
	4. Purge until all parts are free of contaminants and solvents.	Clean, dry nitrogen BB-N-411 at ambient or higher temperature.
Personnel oxygen breathing hose	1. Remove hose from any assembly.	...
	2. Wash hose in soap and water.	Soap, P-S-600 1-oz soap/gal of water or detergent, MIL-D-16791, Type I, 1/4 to 1/2 oz/gal of water.
	3. Clean with soft bristle tube brush.	
	4. Agitate solution around hose.	
	5. Flush with water.	Distilled or demineralized water.
	6. Dry hose thoroughly.	Clean, dry air or nitrogen.
	7. Sanitize the hose by rinsing with a disinfectant.	1:1000 solution of clear merthiolate and distilled water or solution of 0.25 drops of a 10% concentration of benzalkonium chloride and distilled water.
	8. Rinse.	
	9. Dry.	Clean, dry air or nitrogen

NOTE: $t°C = t°F - 32/1.8$. 1 psig = 6.894.757 Pa · s. 1 oz/gal = 7.489 kg/m³.

TABLE 2—*Military and industry survey cleaning methods of oxygen equipment.*

Document	Cleaning Solvents	Material Cleaned As Mentioned	Drying and Purging Gases
MIL-STD-1359A	Trichlorotrifluoroethane MIL-C-81302, Type I or II	All metals, Teflon®	Compressed air BB-A-1034 Technical nitrogen BB-N-411, aviator's oxygen MIL-O-27210
	Laundry soap P-S-600		
	Isopropyl Alcohol TT-I-735	Rubber, metal, gaskets, silicone Hard plastics, metals	
	Nonionic detergent MIL-D-16791	Rubber, metal, gaskets, silicone	
	Chlorinated solvents	Not recommended for use as they are toxic and dangerous.	
USAF Technical Order T.O. 15X-1-1 Section IX (same as above but add following)	Methyl chloroform O-T-620	All metals, Teflon®	same as above
	Trichloroethylene MIL-T-27602	All metals, Teflon®	
Society of Automotive Engineers, Inc. Aerospace Information Report AIR1176	Trichlorotrifluoroethane MIL-C-81302, Type I or II	Metal	Oil-free, dry air or nitrogen filtered thru 5-μm filter
	Nonionic detergent MIL-D-16791	Metal, non-metals[a]	
	Trichloroethane MIL-T-81533A	Metal	
	Isopropyl alcohol Freon TF or Freon 113	Not recommended for nonmetallic articles as the solvent may not flush. Mentioned for Freon compatible materials only.	...
Compressed Gas Association, Inc. Pamphlet G-4.1 "Cleaning Equipment for Oxygen Service"	Methylene chloride	Nylon, polytetrafluoroethylene, glass, and metals	not mentioned
	Refrigerant chloride	same as above	
	Perchloroethylene	same as above	
	Methyl chloroform	same as above	
	Trichloroethylene	same as above	
	Carbon tetrachloride	Use only if absolutely necessary, high toxicity.	

McDonnell Douglas Process Specification 17009.7 "Oxygen System Components; Cleaning and Handling of"	McDonnell Douglas solvent Process Specification 12020 Process Specification 12050.2 Soap, Ivory or Lux Liquid	All metal tubing (except copper) metal fittings, metal heat exchangers exclude metal braided LOX hoses. Copper tubing Fabric covered rubber oxygen breathing hose	Nitrogen BB-N-411, Type I, Class I, Grade A or B, propellant nitrogen MIL-P-27401, passed through a 10-μm filter.
Northrop Process Specifications C-1.7 "Cleaning and Protection of Fluid System Components" and MA-40 "Aircraft Oxygen Systems" Inspection Test Work Order 992592 "Oxygen System Purging Requirements"	Trichlorotrifluoroethane MIL-C-81302 Methyl chloroform MIL-T-81533, O-T-620 Freon TF MIL-C-81302 Dry cleaning solvent, Type I, P-D-680	Metal and compatible nonmetal oxygen system components Referenced in document but application not given Metal and compatible nonmetal oxygen system components Used for plastic and nonmetallic closures not safe for other solvents	Nitrogen BB-N-411, Class I Grade A (Drying clean components) Oxygen MIL-0-27210 (purging aircraft system)
Grumman Aerospace Corp. Manufacturing Engineering Process Standard 7000-70 "Cleaning of Breathing Oxygen System Components"	Trichlorotrifluoroethane (Freon TF or equivalent)	Component free of elastomeric materials (that is, O rings, anti-sieze tape). Only nonmetallic details acceptable per MIL-D-19326 and all other nonmetallics removed (for example, lubricants). Before this cleaning, clean per—MEPS-7000-81 acid cleaning of aluminum; MEPS-7000-82 acid cleaning of ferrous alloys; MEPS-7000-83 vapor degreasing of ferrous and nonferrous alloys (MIL-S-5002) (except titanium) MEPS-7000-84 safety solvents cleaning (except titanium), MEPS-7000-89 ultrasonic cleaning	Compressed nitrogen gas MIL-P-27401 Grumman MSFC-SPEC-234 filtered from a 15-μm filter maximum temperature 150°F except coated aluminum alloy parts maximum temperature 125°F

TABLE 2—(Continued).

Document	Cleaning Solvents	Material Cleaned As Mentioned	Drying and Purging Gases
American Safety Flight Systems, Inc. Process Specification 103, Rev L	Perchloroethylene	All metallic oxygen system components shall be cleaned by vapor degreasing. All nonmetallic oxygen system components shall be cleaned by ultrasonic cleaning	Nitrogen gas per MIL-P-27401 air, filtered, dry and oilfree or by evacuation oxygen per MIL-O-27210
	8 oz of Tyrco 4215 cleaner per 7 gal of tap water in cleaner tank, de-ionized water (7 gal) in ultrasonic rinse tank, heat tanks to 100/150°F. A detailed procedure is described for passivating corrosion resistant steels	300 Series CRES 17 to 4 pH Steel 17 to 7 pH Steel	
Warner Robins Air Logistics Center, Robins AFB, Georgia Cylinders cleaned in accordance with T.O. 15X1-3-1-33 vapor degreased with T.O. 15X1-5-3-3 and T.O. 15X1-4-2-12	Vapor degreaser in MIL-C-81302, Type II	Metal oxygen cylinders pressurized to 2150 psig	Oven backed or dried using breathable air, purged with clean, dry, oil-free oxygen per MIL-O-27210, Type I stored and shipped charged with 24- to 50-psig oxygen
Normalair-Garrett Limited Cleaning Specification	Detergent solutions (Stergene, Lissapol N, Teepol Stripalene 725, Quadralone). 2% sodium cyanide solution in distilled or de-ionized water	Untarnished subassemblies of copper alloys, stainless steels, and mild steels (untarnished). Tarnished copper alloys, tarnished stainless steels, mild steels (rusted and scaled), soldered subassemblies, steel components "sulfinus" treated	Air blast before oven drying maximum temperature of 90°C air is clean and has no dust or oil particles

			Nitrogen per BB-N-411 or oxygen per MIL-O-27210
Essex Cryogenics of Missouri, Inc. document, "General Cleaning and Handling Procedures for Oxygen Components"	Isopropyl alcohol	Nylon, copper alloys (tarnished), stainless steels (scaled and tarnished), mild steels (rusted and scaled), soldered subassemblies, and steel components "sulfinus" treated	
	Toluene	Soldered subassemblies (incorporating Resin Flux Removal)	
	Trichlorotrifluoroethene	Electrical components and printed circuit boards	
	Washing in acetone or mild soap solution followed by hot water rinse	nonmetallic parts	
	Degrease (solvent not given), passivate (method not given), wash in acetone or hot alkaline cleaner followed by hot water rinse	stainless steel parts	
	Degrease (solvent not given), wash in acetone	aluminum parts	

NOTE: This information has been determined from Refs 7 through 21. $t°C = t°F − 32/1.8$. 1 psig = 6894.757 Pa · s. 1 oz/gal = 7.489 kg/m³.
"Nonmetals are synthetic rubber, natural rubber, nylon, polyvinyl chloride fabric, Teflon®, Kel F, polymides, phenolics, and Viton.

ever, a visual inspection is a rapid and inexpensive technique of checking for gross contamination. Parts that are detected as contaminated by visual inspection may be returned to the cleaning solvent solution before oven drying, but certainly before alternate more difficult and time consuming inspections are accomplished.

The most common visual inspection methods are given as follows:

1. Unaided visual inspection: An unaided visual inspection should be conducted in well lighted conditions with all accessible surfaces that would be in contact with oxygen service showing no evidence of rust, dirt, scale, paint, organic materials, grease, oils, ink, and dye. Seams, joints, threads, outlets, and inaccessible portions of the oxygen components may be wiped with clean paper and then inspected. Magnification techniques may also be used to assist in detection of contaminants.

2. Ultraviolet or black light visual inspection: The black light test is another form of a visual inspection, which uses ultraviolet light with a wavelength range of 360 to 290 nm. This technique causes some hydrocarbon contaminants to fluoresce and be more easily detected visually. It is desirable to shield other wavelengths of light from the part under inspection to aid in visual detection of the fluorescing hydrocarbons. A drawback to the black light test is that not all contaminants easily fluoresce. Filter paper may be used to wipe inaccessible areas. Check the filter paper before and after for evidence of the degree of fluorescence.

3. Water break test: Use clean distilled water on the surface of the part inspected. After at least 15 s, the surface may be inspected. The liquid will remain as a film and not form drops on an oil or grease free or clean surface.

4. Acidity or alkalinity pH test: The surface may be wetted with a little distilled water and after 15 s wiped with pH paper. The pH paper should indicate a neutral level of acidity or pH 7 with a small tolerance for paper error pH 6 to 8.

Circumstances may require alternate tests to determine that the desired cleanliness levels have been achieved.

There will be surfaces on oxygen components that are inaccessible to visual inspection. When these surfaces are in contact with oxygen liquid or gas when in service, then they may be subjected to the nonvolatile residue test or the particulate test or both to verify that the required level of cleanliness is being achieved. In the USAF cleaning facilities, a cleanliness level 300C defined by MIL-STD-1246 is desired. These tests are described as follows:

1. Nonvolatile residue (NVR) test: The clean surfaces are rinsed with trichlorotrifluoroethane or equivalent fluid. The volume of fluid that is used should be 21.5 mL/dm^2, and the total fluid volume used is not less than 150 mL. One-half of the fluid used for the rinse is for residue weight and the remainder for particle count. The residue weight is determined by evaporating the entire fluid content at about 45°C (113°F). The separate containers are

compared such that the residue from the rinse fluid does not exceed that from the clean fluid (the reference) by more than 16.15 mg/m^2 (1.5 m/ft^2). This test is not valid with test areas that are less than 0.07 m^2 (0.75 ft^2).

2. Particulate test: Another method used to determine particle content (the contamination), is filter the rinse fluid through a 1.2-μm membrane filter. The filter is then inspected with magnification equipment and the particles are counted and measured. There should not be more than 20 particles at 100 to 175 μm and none larger than 300 μm/100 mL of rinse fluid.

Other test methods are as follows:

1. Infrared spectrophotometry: It could be desirable to determine if trace amounts of hydrocarbon contaminants are present in the cleaning solvent. This may be accomplished with an infrared spectrophotometry test or an infrared analyzer. The cleanliness level may be determined to a few parts per million of hydrocarbon residue by weight. Cleanliness levels should be within the thresholds as given by Table 3.

2. Halogen test: The standard method that is currently in use to detect the presence of cleaning fluid or solvent residue left behind after the final rinse is with the halogen or halide detector.

Package Requirements and Methods to Prevent Recontamination

When oxygen equipment has been cleaned and verified as meeting the established cleanliness criteria, it is essential that measures be taken to prevent recontamination. Whole organizations have evolved within the military to handle, package, and ship anything from miniature components to large, heavy items. Before the oxygen equipment is submitted to shipment, protective measures must be taken to maintain the cleanliness level desired.

Component openings should be sealed with aluminum or plastic caps/plugs. The plastic material for this application is Teflon®, Kel-F®, and Aclar®. These plastic material tradenames are described by MIL-STD-1475 and MIL-HDBK-407 given by Table 4. Hot dip coating compounds should not be used for any purpose as it will be difficult to completely remove all material on assembly. Rubber or cork stoppers, paper, cloth, and tape are not suitable for use as they will leave organic deposits that could later initiate an explosion or fire when the oxygen system is assembled and in service. No preservative of any type should be applied to oxygen equipment. All openings must be sealed, and where plugs or caps cannot be used, a suitable plastic bag shall be used with a twist on. Should items be reassembled in the clean room, then protective plugs, caps, and packages need only be used on the assembled equipment openings.

Small components and parts not exceeding about 77.4 cm^2 (12 in.^{2}) or 4.5 kg (10 lb) may be wrapped with or sealed in bags made of Teflon®, Kel-F®, Alcar®, or fluorohalocarbon of at least 0.04 mm (2 mil) thickness. Other plas-

TABLE 3—*Comparison of purity requirements for breathing and purging gases and liquids.*

Document Constituent	Aviator's Breathing Oxygen MIL-O-27210E Type I, Gaseous	Aviator's Breathing Oxygen MIL-O-27210E Type II, LOX	Aviator's Breathing Oxygen Purity Standard, SAE As 8010 Type I, Gaseous	Aviator's Breathing Oxygen Purity Standard, SAE As 8010 Type II, LOX	Air, Compressed, Breathing Grades A and C	Air, Compressed, Breathing Grades B and D	Nitrogen, Technical Type I, BB-N-411C Grade A	Nitrogen, Technical Type I, BB-N-411C Grade B	Nitrogen, Technical Type I, BB-N-411C Grade C
Oxygen, %	>99.5	>99.5	>99.5	>99.5	20 to 22	19 to 23	0.05	<0.50	0.50
Nitrogen, %	<0.5	<0.5	<0.5	<0.5	...	...	99.5	>99.5	99.5
Carbon dioxide (CO_2), ppm	10	5	10	5	500	1000	...	...	...
Carbon monoxide (CO), ppm	...	...	...	...	10	10	...	...	...
Methane (CH_4), ppm	50	25	50	25	25	...	50	50	50
Acetylene (C_2H_2), ppm	0.1	0.05	0.1	0.05	...	...	...	...	...
Ethylene (C_2H_4), ppm	0.4	0.2	0.4	0.2	...	...	...	...	...
Ethane (C_2H_6) and hydrocarbons, ppm	6	3	6	3	25	...	50	50	50
Nitrous oxide (N_2O), ppm	4	2	4	2	...	...	...	...	...
Solvents, halogenated, ppm	0.2	0.1	0.2	0.1	0.2	...	...	...	...
Refrigerants, halogenated, ppm	2	1	2	1	0.2	...	...	...	...
Other compounds, ppm	0.2	0.1	0.2	0.1	0.02	...	...	...	...
Moisture, mg/L	0.005	0.005	0.005	0.005	0.02	0.3	0.02	0.02	...
Solid particles, μm	...	<1000	<100	<1000	...	...	<50	<50	<50
Solid fibers, μm	...	<6000	<600	<6000	...	...	...	...	...
Solids, density	...	1 mg/L	1 mg/m³	1 mg L	...	...	...	...	...
Oil and particulate matter, mg/L	...	...	...	...	0.005	0.005	free	free	free
Odors	none	none	none	none	not objectionable	not objectionable	none	none	none

NOTE: This information has been determined from Refs 22 through 25.

TABLE 4—*Military documents that are applicable to packaging requirements.*

Document	Application
MIL-STD-1475	Contamination Control Technology Overview of Precision Cleaned Material
MIL-STD-1246	Product Cleanliness Levels and Contamination Control Program
L-P-378	Plastic Sheet and Strip, Thin Guage, Polyolefin
MIL-B-117	Interior Packaging Sleeve and Tube Bag
MIL-C-5501	General Specification for Dust and Moisture Seal Protective Caps and Plugs
MIL-STD-129	Marking for Shipment and Storage
MIL-STD-794	Procedures for Packaging and Packing of Parts and Equipment
MIL-P-116	Methods of Preservation
MIL-HDBK-304	Package Cushion Design
MIL-HDBK-407	Contamination Control Technology
MSFC-SPEC-456	Fluorohalocarbon Film, 0.05 μm (2 mil)

tic bags, such as polyethylene, may only be used as another outer protective covering and may not contact clean oxygen components. Note that the packaging material that comes in contact with the clean components must be clean to the same cleanliness level as the oxygen components to be packaged. Purging with clean, oil-free, dry gas may be used to clean the packaging as solvents, and cleaning materials will be very difficult to remove. Heat sealing is preferred, but staples and tape may be used provided the open end is doubled over first. Cushioning material must be used to protect seals while the article is in a box in shipment. Outer packaging is a must for long-term storage or whenever there is a probability the seal may be broken on the inner bag.

Tags must be securely attached to each packaged item to identify that the component is cleaned. It is standard practice in the military to mark the identification tag with the component name, "Life Support," equipment label, part number, and one of the following: "OXYGEN CLEANED," "CLEANED FOR OXYGEN SERVICE," or "ALL OIL, GREASE, SHOP RESIDUE, OR OTHER CONTAMINANTS HAVE BEEN REMOVED. DO NOT OPEN UNTIL READY FOR USE."

Cleaning Oxygen Components and Plumbing on the Military Aircraft

A thorough discussion has been provided that elaborates cleanliness levels, methods, tests, inspections, and packaging for oxygen equipment within a controlled environment or clean room. In many ways, cleaning oxygen equipment within a clean room is far easier than at the aircraft. In the clean room, direct visual inspections may be used the majority of the time to check cleanliness levels. Visual inspections of most surfaces that are in direct contact with oxygen service are easier because most components have been broken down to the smallest piece. The exception may be containers such as LOX converters.

Direct visual inspection techniques are more reliable when component contamination must be determined.

Military aircraft gaseous and liquid oxygen systems are closed to the ambient after the installation is complete. Should particle, hydrocarbon, cleaning solvent, or moisture contamination be included within the oxygen components or plumbing or both, measures must be taken to clean the system. It may not be necessary to break down the system installation if particles are small enough to be flushed through the lines. The two cleaning methods that may be used are solvent flushing and then hot gas purging or strictly hot gas purging. Typically, it will not be easy to ascertain that solvent flushing is necessary. Generally, solvent flushing at the aircraft is only performed on a new or overhauled oxygen system installation. Solvent flushing is most effective in removing particles and hydrocarbons. Purging should be performed whenever the lines have been depressurized for a minimum of one 24-h daily cycle. Ambient barometric changes are likely to bring into depressurized oxygen lines airborne contaminants that must be purged.

Hot gas purging is most effective in removing moisture and solvents but is not effective in removing particles and hydrocarbons from the system. Particles may be effectively removed by sonic hot gas purging, but this has never been used on past aircraft systems because of valve restrictions and purge equipment limitations.

Remaining solvent and hydrocarbon type contamination may be determined by merely pressurizing the lines with clean oil-free oxygen and breathing 100% oxygen from the system. Solvent and hydrocarbon contamination in excess of that allowed for oxygen gas and liquid (see Table 3) will nearly always be detectable as an odor. Extreme caution should be used in any sniff test that is used. The person should not breathe 100% of the potentially contaminated gas, only for a short time period (sniff test). If gross contamination is detected, there is the potential for an extreme fire or explosion and the system must be cleaned immediately. Oxygen should first be removed from the lines.

Particle and moisture contamination will not be easy to determine. The results may be catastrophic when a line component fails from this contamination. In this event, oxygen supply to the aircraft occupants will probably be precluded. When orifices and valves are clogged with solid or ice particle contaminants, regulation and delivery of oxygen supply may be restricted. Should a LOX converter pressure relief valve/device be restricted, then an overpressurization may occur. If the converter is more than one-half full, a hydraulic failure is likely, resulting in a rupture of the side walls and LOX spillage outward. If the converter is less than one-half full, then a pneumatic failure is likely to occur. Such a violent explosion is likely to damage the aircraft. Past records show that both types of incidents do occur infrequently on military aircraft.

The best approach to minimize moisture contamination is to implement periodic hot gas purges to the aircraft oxygen system. The frequency will be a function of the type of oxygen system installation (gaseous, permanent LOX converter versus removal LOX converter).

Flushing and purging aircraft oxygen system components are best accomplished by cleaning segments of the system. For example, current methods of purging the converter are to disconnect the supply line, connect the purge unit with appropriate gas supply to the supply line (fill valve), and purge at 383 to 394 K (230 to 250°F) at a minimum of 207 to 345 kPa (30 to 50 psig) for about 1 h. The outlet temperature should reach a minimum of 352.4 K (175°F) for effective purging. Components and plumbing of the aircraft may also be purged in the same way with segments purged alternately. Open clean lines should be covered with caps or suitable plastic bags tied on with twist ons. Any article that may later become Foreign Object Damage (FOD) for the aircraft engine should not be used.

Nonvolatile freon cleaning solvents are best for flushing. Any solvent agents that when mixed with 394 K (250°F), 207 to 345 kPa (30 to 50 psig) oxygen gas that would be a fire or explosion hazard should not be used. Viscous freon fluids will carry out solid particles when flushed through, and some freons also have an affinity for water moisture. Gas purging will not have enough viscosity to carry through larger solid particles.

There has been much debate over the safest and most effective hot gas purging method. Past procedures allowed hot clean nitrogen gas purging at the aircraft since heated nitrogen gas poses very little fire and explosion hazard. It was also thought that nitrogen gas had more affinity for water moisture than did oxygen gas. The primary problem with nitrogen gas purging was the risk that some nitrogen gas may be trapped or held in the system and not be flushed out in a later oxygen purge. Also, the maintenance crew may neglect to finish with an oxygen purge for some reason or another. High concentrations of nitrogen gas are hazardous to the person breathing such gas and he will rapidly lose consciousness. An exception to this would be that a nitrogen purge is much safer to use when it is desired to purge after a solvent flushing. However, an air or oxygen purge must always be used after a nitrogen purge at the aircraft no matter the reason. In all cases, only clean, oil and moisture free nitrogen, air, and oxygen gas should be used for purging. The purging unit should also incorporate a 5 μm nominal rated filter to preclude entry of solid particles into the aircraft oxygen system. The cleanliness criteria of all these gases are defined for air, BB-N-411C, Type I or II, Grade A or B, Class I only (Class II may be oil pumped), and oxygen MIL-O-27210E, and SAE AS 8010, Types I and II.

Rationale for Cleaning Military Aircraft Oxygen Systems

Cleanliness requirements for aircraft oxygen systems have been undergoing a continuous evolution. As knowledge of oxygen and contaminant interaction has advanced new and improved cleaning methods, and inspections have been determined. A contaminant is defined as any material that by being present in a system or component may cause mechanical malfunction, fire, or explosion. The goal in cleaning military aircraft oxygen systems is to remove contaminants that would

1. Result in noxious odors adverse to those breathing from the source of supply and delivery components.
2. Preclude effective operation of movable components for proper regulation and delivery of oxygen supply.
3. Preclude effective operation of the oxygen system because small openings and orifices are clogged or restricted.
4. Minimize fire and explosion hazards from contaminants that may easily ignite in the presence of 100% pressurized oxygen gas or liquid.
5. Minimize fine metallic or nonmetallic particles that would ignite in high velocity 100% oxygen gas after impingement of materials with a low threshold of ignition.

Generally, a clean oil-free gas that meets the limits given for gases in Table 3 will not have odors that will be noxious to any aviator's breathing. The human sense of smell is reportedly the best odor detector available and very minute amounts of contaminants are easily detectable. Of particular concern is that any cleaning solvents are not still at surfaces in contact with oxygen service. Proper hot purging will in time eliminate these odors.

Solid, liquid, and grease like contaminants may jam or gum up movable mechanisms and clog up openings in oxygen components. The most commonly expected problem is that water moisture will condense on valves and components as the aircraft ascends to higher altitudes. Below freezing temperatures result in stoppage of moving mechanisms contaminated with moisture.

The possibility of a fire or explosion is a much more complicated phenomenon not easily predicted [27–30]. Some of the factors involved are the materials, the line pressure and temperature, properties of the oxidizer, and the ignition event. These properties are discussed in detail by Lowrie, Werley, Monroe, and Slusser in ASTM STP 812 [29].

A simple gas dynamics model is given here to describe how a simple chain of events may lead to an oxygen systems fire. This example is illustrative of conditions that may be expected in maintenance and operation of the aircraft oxygen system.

The compression of oxygen gas involves a significant amount of energy that raises the gas and surrounding component temperatures. In the event of a rapid compression, the heat generated will cause a sharp rise in localized temperature. For an adiabatic isentropic compression the temperature rise is given by the formula [31]

$$\left(\frac{T'}{T_o}\right) = \left(\frac{P}{P_o}\right)^{\frac{k-1}{k}}$$

where

To = the initial absolute temperature, °R,
T = the final absolute temperature rise, °R,
K = the ratio of specific heats = 1.4,
Po = the ambient absolute pressure, psi, and
P = the supply absolute pressure, psi.

For example, for rapid compression from atmospheric pressure to a gage pressure of 1146 MPa (1850 psi)

$$(T + 460°F)/(70°F + 460°F) = [(1850)/(14.7)]^{0.286}$$

$$T = 1657°F (903°C)$$

The temperature rise alone is sufficient to cause ignition of organic materials, such as oil, grease, solvents, dust, lint, and many inorganic materials such as metal particles and metal oxide particles. This type of temperature rise is expected in the filling or servicing of pressure vessels. Also, pressure rises in line valves and regulators will elevate the gas temperature. Oxygen flowing through valves and lines may propel particles with such velocity that the friction and momentum of their impact releases enough additional energy to exceed the heat of reaction of the particle. At the elevated temperatures from adiabatic compression, less additional energy is needed to initiate combustion. Low energy ignition can lead to higher heats of reaction igniting metal sections. This is particularly true for aluminum that burns most rapidly among metals and releases significant amounts of energy on combustion. Table 5 illustrates a comparison of materials commonly used in military oxygen systems.

TABLE 5—*Ignition properties of materials commonly used in oxygen service.*

Material	Ignition Temperature		Heat of Combustion, J/g
	°F	K	
Aluminum	1341	1000	31 000
Copper	1881	1300	2 400
Brass	1832	1273	3 600
Inconel	2457	1620	4 700
Steels	2061 to 2547	1400 to 1670	7 500 to 8 000
Nylon	396	475	7 000
Teflon®	801	700	1 100

NOTE: This information has been determined from Refs 27 and 28.

Note that aluminum will release heat in a ratio of 31 000/2400 = 12.92 that of copper and 31 000/8000 = 3.875 that of steels for the same order of magnitude of ignition temperature. Nylon will ignite within the temperature range of adiabatic compressive temperature rise.

The ignition probability will also be aggravated by higher pressures in the lines versus ambient pressure at which the table was determined. It can easily be seen that the oxygen systems design has little margin for safety where contamination is concerned.

Conclusions

It has been determined that cleanliness of military aircraft oxygen systems is critical to preclude component line failures, eliminate noxious odors, and minimize the risk of a fire or explosion to an acceptable hazard level. Many precautions are taken in industry and military organizations to clean oxygen systems and equipment. Most methods and techniques have evolved from considerable background experience relative to cost effective methods of cleaning, solvent flushing, purging, inspecting and testing cleanliness, and packaging oxygen equipment. The application of recent surveys, tests, and studies may provide low cost improvements to cleanliness criteria. Standardization of issues discussed in this paper may enhance confidence in cleanliness criteria. Minimizing problems in the application of the standard cleanliness criteria should be a goal in industry and the military.

References

[1] Schroll, D. W. "Oxygen Standardization Coordinating Group Meeting No. 40 Minutes," Battle Creek, MI, June 1983.
[2] Schroll, D. W., "Oxygen Standardization Coordinating Group Meeting No. 41 Minutes," Tucson, AZ, 25 Nov. 1983.
[3] Schroll, D. W., "Oxygen Standardization Coordinating Group Meeting No. 42 Minutes," Wright-Patterson AFB, OH, 7 June 1984.
[4] Schroll, D. W., "Oxygen Standardization Coordinating Group Meeting No. 43 Minutes," Buena Park, CA, 30 Nov. 1984.
[5] Bankaitis, H. and Schueller, C. F., *ASRDI Oxygen Technology Survey, Volume II: Cleaning Requirements, Procedures, and Verification Techniques*, NASA Special Publication NASA SP-3072, Aerospace Safety Research and Data Institute, Lewis Research Center, Cleveland, OH, 1972.
[6] "Environment Control Procedure, Life Support Equipment General Requirements for Cleaning, Inspection, Packaging and Preservation," U.S.A.F. Technical Order T.O. 15X-1-1, Section IX, Aug. 1982.
[7] Olsen, W., "Application of MIL-STD-1359A to Military Oxygen Equipment," letter, Dayton T. Brown Inc., New York, 21 July 1983.
[8] Nakagiri, K., "MIL-STD-1359A", Lockheed Aircraft California letter, 26 July 1983.
[9] Johnston, V. P., "Oxygen System Cleaning and Purging Practices," letter McDonnell Douglas Corporation, St. Louis, MO, 7 Oct. 1983.
[10] Majeaky, R. L., "Cleaning and Handling of Oxygen System Components," Process Specification 17009.7, McDonnell Douglas Corporation, St. Louis, MO, 9 March 1984.
[11] Torgerson, R. E., "Cleaning and Protection of Fluid System Components," Process Specification C-1.7, 30 Oct. 1979; "Aircraft Oxygen System," Process Specification MA-40, 15

July 1975; and "Oxygen Installation on F-5F Aircraft," Inspection Test Work Order 992592, Northrop Corporation, CA, 10 July 1974.

[12] Sheehan, J. J., "MIL-STD-1359A Evaluation," letter, McDonnell Douglas Corporation, St. Louis, MO, 29 Sept. 1983.

[13] Tallman, B., "Oxygen Degreasing Procedure Used Years Ago," Process Standard 7000-70, "Cleaning of Breathing Oxygen System Components," 13 Nov. 1972, "Oxygen System Cleaning Procedure F-14 Aircraft," Grumman Aerospace Corp., New York, 7 March 1983.

[14] Barmasse, A. C. and Sloan, J. J., "Cleaning Procedure" American Safety Flight Systems," Process Specification No. 103, CA, 26 July 1983.

[15] Rybicki, J., "Cleaning Methods and Procedures for Breathing Oxygen Equipment," letter, Warner Robins Air Logistics Center, Robins AFB, GA, 8 Aug. 1983.

[16] Browne, P. G., "Cleaning of Oxygen Breathing System Parts," Specification, Normalair-Garrett Limited, Phoenix, AZ, 15 Aug. 1983.

[17] Martin, A. and Czyzewski, C., "Information MIL-STD-1359A," letter, Naval Air Engineering Center, Lakehurst, NJ, 25 Aug. 1983.

[18] Geren, D., "General Cleaning and Handling Procedures for O_2 Components," Essex Cryogenics Cleaning Procedures GP-19, Essex Cryogenics of Missouri, Inc., St. Louis, MO, 30 Aug. 1983.

[19] "Cleaning Equipment for Oxygen Service," CGA, Pamphlet G-4.1, Compressed Gas Association, Inc., Arlington, VA, 1977.

[20] Ott, W., "MIL-STD-1359A Review," memo, 24 Aug. 1983, "MIL-STD-1359A Review," memo, 15 Jan. 1976; and Standard ST 1637806 "Cleanliness Requirements for Oxygen Equipment," SP 1636959, "Specification for Cleanliness Requirements for Oxygen Equipment," 3 March 1982, and Standard PD 1635249, "Guidelines for Use by Process Engineers in the Selection of Cleaning Procedures," 18 Jan. 1973, Clifton Precision, Davenport, IW.

[21] Society of Automotive Engineers, Inc. (SAE) Committee A-10, Aircraft Oxygen Equipment, "Oxygen System and Component Cleaning and Packaging," Aerospace Information Report AIR 1176, Aug. 1974.

[22] Society of Automotive Engineers, Inc. SAE Committee A-10 Aircraft Oxygen Equipment, "Aviators Breathing Oxygen Purity Standard," Air Standard AS 8010, 2 Oct. 1984.

[23] Military Specification MIL-O-27210E, "Liquid and Gas Aviator's Breathing Oxygen," Department of Defense Document, San Antonio ALC/SFRM, Kelly Air Force Base, TX, 28 April 1980.

[24] "Compressed Air for Breathing Purposes," Federal Specification BB-A-1034A, Federal Agency Document, 15 Dec. 1970.

[25] "Technical Nitrogen, Gaseous and Liquid," Federal Specification BB-N-411C, Federal Agency Document, Department of the Army, AMSME-RZD Fort Belvoir, VA, 3 Jan. 1973.

[26] Yuen, H. H. and Sander, L. F., "Compatibility of Materials with Oxygen," Naval Air Engineering Center, NAEC MISC 92-0354, Lakehurst, NJ, 9 Sept. 1978.

[27] Kuchta, J. M., "Fire and Explosion Manual for Aircraft Accident Investigators," Technical Report AFAPL-TR-73-74, Air Force Aero Propulsion Laboratory, Wright-Patterson Air Force Base, OH, Aug. 1973.

[28] Bond, A. C., Pohl, H. O., Chaffee, N. H., and Stradling, J. S., "Design Guide for High Pressure Oxygen Systems," NASA Reference Publication 1113, National Aeronautics and Space Administration, 1983.

[29] Werley, B. L., Ed., *Flammability and Sensitivity of Materials in Oxygen Enriched Atmospheres, STP 812*, American Society for Testing and Materials, Philadelphia, 1983.

[30] Society of Automotive Engineers, Inc. SAE Committee A-10, Aircraft Oxygen Equipment, "Transfilling and Maintenance of Oxygen Cylinders," Aerospace Information Report AIR 1059A, 15 Oct. 1982.

[31] Shapiro, A. H., "The Dynamics and Thermodynamics of Compressible Fluid Flow, Vol I," The Ronald Press Company, New York, 1953.

John A. Gilbertson[1] *and Robert Lowrie*[2]

Threshold Sensitivities of Tests to Detect Oil Film Contamination in Oxygen Equipment

REFERENCE: Gilbertson, J. A. and Lowrie, R., **"Threshold Sensitivities of Tests to Detect Oil Film Contamination in Oxygen Equipment,"** *Flammability and Sensitivity of Materials in Oxygen-Enriched Atmospheres: Second Volume, ASTM STP 910*, M. A. Benning, Ed., American Society for Testing and Materials, Philadelphia, 1986, pp. 204–211.

ABSTRACT: Equipment and components for use in an oxygen service are required to be "clean" or "free from oil," as contamination by oil is believed to have been responsible for numerous incidents. In some specifications, cleanness is defined by stating the maximum allowable concentration in grams per square metre, but often the specification is qualitative. Four tests of cleanness are widely quoted; these are visual examination in white and in ultraviolet light, a wipe test, and a water break test. The threshold sensitivities of these four tests have been evaluated for six different hydrocarbon oils commonly used in the industry. The results of this limited project show that under the test conditions the order of increasing sensitivity was white light, ultraviolet (UV) light, modified wipe, and water break. Water break was 20 to 40 times more sensitive than the white light test. Considerable further investigation is recommended.

KEY WORDS: oxygen, equipment, oils, detection, tests, safety, cleanness, contamination, cleanliness

Experts in the production and handling of oxygen are convinced that systems contaminated with oil are hazardous and are more or less susceptible to fire depending on the conditions. Many data have been collected on the amount of oil contamination likely to cause an ignition, and it is known that appropriate cleaning procedures are capable of achieving cleanness levels lower than the regions of oil flammability. Werley [1] reports: "verification (of cleanliness) has required a mix of direct and indirect examination meth-

[1]Group technical manager-Gases, The BOC Group Plc, Hammersmith House, London W6 9DX.
[2]Senior research metallurgist, The BOC Group Inc., Group Technical Center, Murray Hill, NJ 07974.

ods. Each of these had its own limitations; however, in combination the result is achieved, not easily, but realistically."

We have evaluated the threshold sensitivities of four tests commonly used in the industrial gases industry for the detection of oil contamination [2]. The tests were as follows:

- White light examination
- Ultraviolet light examination
- Modified wipe test
- Water break test

The tests were applied to six different oils commonly used in the United Kingdom.

Test Procedures

Preparation of Test Panels

Stainless steel panels of dimensions 153 by 100 by 6 mm were prepared and polished to a surface finish of 0.75 μm. The panels were cleaned before oil application by brushing with detergent solution followed by rinsing with water, distilled water, acetone, and finally carbon tetrachloride. The plates were then stored in a desiccator. The six hydrocarbon oils selected are described in Table 1. They include lubricants for a variety of situations.

TABLE 1—*Description of oils tested.*

Oil	Supplier	Type	Application
Mobil Rarus 57		mineral oil/ synthetic product blend	air compressor cylinder lubricant
Mobil DTE oil light		high viscosity mineral oils blended with antioxident and anti-rust additives	turbo compressor bearing lubricant
DTE medium	Mobil		expansion turbine lubricant
Mobil DTE oil medium heavy			crankcase lubricant
Gargoyle arctic oil light		low viscosity mineral oil	low-temperature oil for cylinder and bearing lubrication
Edwards high vacuum oil grade 18	Edwards High Vacuum	mineral oil blended with multifunctional additives	mechanical vacuum pump lubrication particularly where corrosive gases/ vapors may be encountered

Several methods of coating surfaces with a uniform oil film at known concentration were investigated. Depositions of oil films by evaporation from solvent solutions containing known quantities of oil were generally unsatisfactory. Evaporation of known volumes of standard oil solutions in a variety of solvents on metal panels and trays failed to give films of uniform thickness. The oil tended to concentrate either in the center of the tray or at the edge depending on the solvent used. Application using a rubber faced printing roller was also abandoned because the method could only produce thick films relative to those required. This method was also rejected because of the possibility of contamination by additives from the polymer surface of the roller, which might have influenced the ultraviolet (UV) or water break tests. Other coating techniques used in the paint industry for the preparation of thin films were also considered. These techniques used mechanical apparatus to spray plates or withdraw panels at uniform rates from tanks of paint. These methods were rejected because they were either too laborious or required the construction of sophisticated and expensive equipment. The technique finally adopted was to spray the panels with a solvent solution of the oil using a hand-held compressed-air-driven paint spray gun moved across the plate. Oil solutions at concentrations of 0.1% v/v and 1% v/v were prepared in a solvent mixture consisting of

(1) petroleum spirit (boiling range 60 to 80°C), 25 parts by volume and
(2) industrial methylated spirit (ethanol denatured with methanol), 75 parts by volume.

Solvent grades with low residue on evaporation were chosen. The paint sprayer used was a Colospray Model CGX manufactured by Coloursprays Ltd, London.

Four of the test panels were assembled in a block 306 by 200 mm on an inclined surface facing the spray gun. The plates were sprayed by repeated sweeps with the gun, and after allowing the solvent to evaporate, the plates were examined for activity by the appropriate test.

Detection Procedures

The examination for activity in each test was made by comparison with a "blank" panel that had been cleaned as previously described but had not been sprayed with oil solution. The examination was made by at least three observers on a "blind" basis. Four test panels plus the blank were observed simultaneously, but the identity of the blank plate was not disclosed to the observer. The spraying and observation procedure was continued until a majority of observers detected activity using the appropriate test. Plates were similarly sprayed with pure solvent. These plates showed no activity in any of the tests.

Visual Light Test

The panels were examined visually in a north-facing room illuminated by fluorescent lighting.

Ultraviolet Test

The panels were examined in the darkened room but not total darkness. The panels were illuminated using a Hanovia UV/fluorescence lamp, wavelength 150 to 350 nm held 30 cm from the panels at an angle of approximately 45°. Observations were made along the line of incident light after allowing the observers' eyes to adjust to the reduced light level for several minutes.

Wipe Test

In practice, the wipe test is normally carried out by wiping an area of surface approximately 10 to 20 cm² with a glass fiber filter paper wrapped around a finger, followed by a visual examination of the paper for staining. This method was found to be unsatisfactory. Even new panels, cleaned as previously described and oil-free, produced a visual stain, probably because of traces of oxides or even abraded metal. As a result, it was decided to combine the wipe test with an UV examination of the paper. A Whatman GF/C glass-fiber paper was held over the finger, which was then drawn vertically across the surface wiping a strip approximately 15 cm long by 1 cm wide. The paper was then examined for fluorescence under ultraviolet light.

Water Break Test

The horizontal surface was wetted with a fine spray of distilled water, which formed a film that remained unbroken, that is, continuous, for several seconds. In the presence of oil, the film rapidly broke up into small beads of water. The behavior of clean and oil-contaminated panels is shown in Fig. 1.

Analysis

After activity had been detected in each test, the oil film was carefully rinsed off with hydrocarbon-free carbon tetrachloride,[3] and the washings were diluted to a known volume in a volumetric flask. For each wipe test, the amount of oil was measured on a second panel sprayed at the same time. The oil content of the solution was then determined by infrared spectroscopy by the following procedure.

---

[3]Because of its toxic and carcinogenic properties, carbon tetrachloride is no longer used for such analytical work at BOC. Freon 113 (1,1,2-trichlorotrifluoroethane) is used instead.

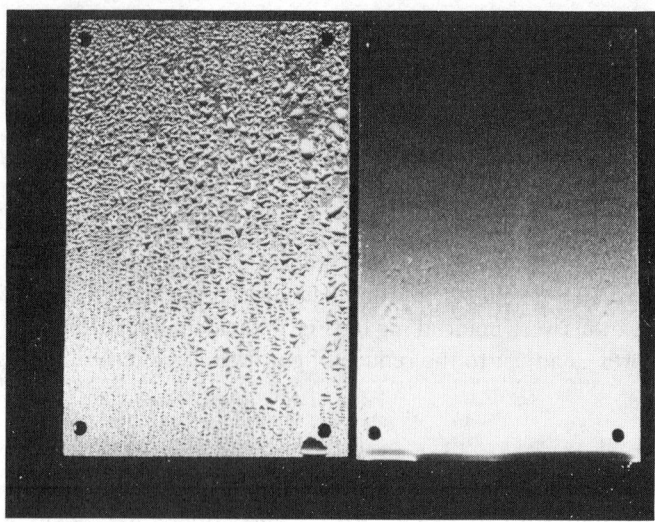

FIG. 1—Right: *an oil-contaminated panel and* Left: *a clean panel after spraying in the water break test.*

The infrared spectrum was scanned using a Perkin Elmer 457 spectrophotometer over the wave number range 2700 to 3100 cm^{-1} (wavelengths of 3.2 to 3.7 μm) and the absorbance at 2930 cm^{-1} (3.4 μm), the carbon-hydrogen (C-H) stretching peak, was measured. The dilution of the solution and the cell path length were chosen so that each solution gave an absorbance of 0.3 to 0.4 absorbance units. Pure carbon tetrachloride[3] was used in the reference cell. The oil content was then determined by reference to a calibration graph prepared from known concentrations of each oil in carbon tetrachloride.

Results and Discussion

Tables 2 through 5 show the mean value of the concentration of oil in milligrams per square metre at the threshold of detection in each test. The standard deviation is also given.

TABLE 2—*Detection thresholds in white light examination.*

Oil	Mean Threshold of Detection, mg/m^2	Standard Deviation
Mobil Rarus 57	969	78
Mobil DTE light	880	45
Mobil DTE medium	1699	169
Mobil DTE heavy medium	1111	93
Gargoyle arctic light	974	94
Edwards high vacuum grade 18	1253	55

TABLE 3—*Detection thresholds in ultraviolet light examination.*

Oil	Mean Threshold of Detection, mg/m^2	Standard Deviation
Mobil Rarus 57	42	2
Mobil DTE light	815	70
Mobil DTE medium	645	111
Mobil DTE heavy medium	1340	81
Gargoyle arctic light	1478	112
Edwards high vacuum grade 18	243	27

TABLE 4—*Detection thresholds in the wipe test.*

Oil	Mean Threshold of Detection, mg/m^2	Standard Deviation
Mobil Rarus 57	28	3
Mobil DTE light	275	28
Mobil DTE medium	156	29
Mobil DTE medium heavy	568	5
Gargoyle arctic light	415	48
Edwards high vacuum grade 18	113	1

TABLE 5—*Detection thresholds in the water break test.*

Oil	Mean Threshold of Detection, mg/m^2	Standard Deviation
Mobil Rarus 57	56	4
Mobil DTE light	49	7
Mobil DTE medium	42	5
Mobil DTE heavy medium	40	10
Gargoyle arctic light	30	4
Edwards high vacuum grade 18	51	6

As can be seen from Table 6, in which these results are summarized, the order of increasing sensitivity is

(1) visible light test,
(2) UV light test,
(3) wipe test,
(4) water break test.

The variation in response between different oils is most marked in the UV test and the modified wipe test. UV fluorescence in oil is caused by impurities or the additives. Rarus 57, which contains a high concentration of synthetic material is particularly active in the test. The responses in the visible light test

TABLE 6—*Summary of results of threshold of detection, mg/m².*[a]

Test	Mobil Rarus 57	Mobil DTE Light	Mobil DTE Medium	Mobil DTE Heavy Medium	Gargoyle Arctic Light	Edwards Grade 18
White light	*969*	*880*	1699	1111	*974*	1253
UV	**42**	*815*	*645*	1340	1478	*243*
Wipe	**28**	*275*	**156**	*568*	*415*	**113**
Water	**56**	**49**	**42**	**40**	**30**	**51**

[a]Numbers in boldface type ≤ 200 mg/m² and those in italic type > 200 mg/m², ≤ 1076 mg/m².

and the water break test are functions of the bulk properties of the oil. Thus, the differences between oil types are less marked.

Variations that might occur in the field, such as intensity of the UV source and variations in additive content in different batches of oil, may influence the effectiveness of the UV test.

There are no universally accepted limits for the permissable amount of hydrocarbon oil contamination on a part for oxygen service. The Industrial Gases Committee of the Commission Permanente Internationale suggested in a draft of a specification [3] that, for service in oxygen above 435 psig, the maximum hydrocarbon amount be 200 mg/m² (19 mg/ft²). The figures in bold type in Table 6 lie below that level. The Compressed Gas Association (CGA) in Pamphlet G-4.1 [2] suggests 1076 mg/m² (100 mg/ft²) as a limit for hydrocarbon contamination. The threshold sensitivities lying below this level but above 200 mg/m² are shown in italics in Table 6. There remain five combinations of oil and test with threshold sensitivities above the CGA limit.

Further Investigation

Clearly, this test program is not definitive. It only indicates the range of behavior to be expected and the types of experiments needed to complete the picture. The examination was limited to hydrocarbon type oils, but it could usefully be extended to animal, vegetable, halocarbon, silicone, and soluble cutting oils. The tests were carried out on a plane surface with uniform surface finish. Further tests could include nonplanar surfaces and variations in surface finish. Corrosion and surface coatings could be investigated.

The scope of this work is limited, but we hope that it will stimulate interest and that complementary experiments will be made.

Acknowledgment

The authors wish to thank the Industrial Gases committee of the Commission Permanente Internationale Européene des Gaz Industriels et du Carbure de Calcium for permission to publish these results of work commissioned by the IGC, and John Sanders, Analytical Services, BOC Ltd., London, who performed it.

References

[1] Werley, B. L. "Oil-Film Hazards in Oxygen Systems," Flammability and Sensitivity of Materials in Oxygen-Enriched Atmospheres, *STP 812*, American Society for Testing and Materials, Philadelphia, 1983, pp. 108–125.
[2] Compressed Gases Association, *Cleaning Equipment for Oxygen Service*, Pamphlet G-4.1, 2nd ed., 1977, p. 10.
[3] Industrial Gases Committee, Oxygen Cleanliness Standard, 2nd Draft, Commission Permanente International, Paris.

Peter Ernst[1]

New Oil Detector System Enables Fully Instrumental Operation of Oxygen Reciprocating Compressors

REFERENCE: Ernst, P., "New Oil Detector System Enables Fully Instrumental Operation of Oxygen Reciprocating Compressors," *Flammability and Sensitivity of Materials in Oxygen-Enriched Atmospheres: Second Volume, ASTM STP 910*, M. A. Benning, Ed., American Society for Testing and Materials, Philadelphia, 1986, pp. 212–219.

ABSTRACT: This paper presents a new electronic oil detector system that guarantees a continuous monitoring of the function of the oil retention system of oxygen reciprocating compressors. As a measuring method the system uses the variation in the capacitance of a capacitor, as a result of absorption of oil by an absorbent element serving as a dielectric. Field experience shows that this system is a major step forward in safe, automatic, remote control of oxygen reciprocating compressors.

KEY WORDS: oxygen, control systems, monitoring systems, capacitors, dielectrics, safety, oilfree reciprocating compressor, oil detectors

A safe automatic remote control of an oxygen reciprocating compressor was formerly not possible. Pressures and temperatures are relatively easy to monitor. Appropriate instruments like pressure and temperature limit switches as well as safety valves are available. Monitoring of the function of the oil retention system to ensure its safe operation is, however, more difficult.

All oilfree reciprocating piston compressors must, to some extent, be designed according to the same fundamental principles, although the sealing system on pistons and glands may be entirely different. A typical feature common to almost all designs is the classical crank gear with a crosshead. As this is an oil-lubricated part, an efficient barrier must be placed between the driving part and the dry cylinder to exclude carryover of oil.

The dry part can be separated from the lubricated part by a distance piece

[1] Manager of project engineering, Sulzer Burckhardt Engineering Works, Ltd., CH-8401 Winterthur, Switzerland.

of a length equivalent to, or slightly longer than the stroke, arranged between a bulkhead in the frame carrying a package of oil-scraper rings and the pressure packing on the cylinder side.

The scraper ring is the most important item as it has to remove the oil from the surface of the oscillating piston rod. Although this cannot be done perfectly, the remaining layer of oil on the piston rod must be sufficiently fine to stick to the rod rather than cluster in drops, which would be heavy enough to be thrown towards the dry-running side. A ring, working as an oil shield, is normally fixed on the piston rod to prevent oil drops entering into the glands. Even with very efficient oil retention systems, monitoring of the condition in an oxygen reciprocating compressor is necessary.

Accidents on oxygen compressors have led to the situation that those machines, irrespective of their origin and quality, are regarded as dangerous objects. In many countries, therefore, in order to protect operators, it has been recommended that protective walls be erected, for example, of concrete, around oxygen machines (Fig. 1). Access to the running machine is not allowed; therefore a visual monitoring by the operator through the openings in the distance piece is no longer possible.

FIG. 1—*Oxygen labyrinth-piston compressors in an air separation plant.*

Different techniques are applied today to monitor the condition in the distance piece:

• Regular machine shutdown. Visual checking and cleaning of the piston rod if necessary, while compressor is shut down.
• Illumination of the distance piece by ultraviolet light. Inspection of the critical section through ports or windows in the protective walls, using, for example, binoculars.

However, all these procedures are for occasional inspections and have the disadvantage of indicating the instantaneous condition only.

A newly developed electronic system allows a continuous and very sensitive detection of oil scraper leakage. Figure 2 illustrates schematically a cross section of an oilfree reciprocating labyrinth-piston compressor. The marked ring shows the area where the oil detector is placed. Figure 3 shows this area of a standard compressor, and Figure 4 shows the same area with the oil detector.

The object of the invention is to provide a monitoring system that detects

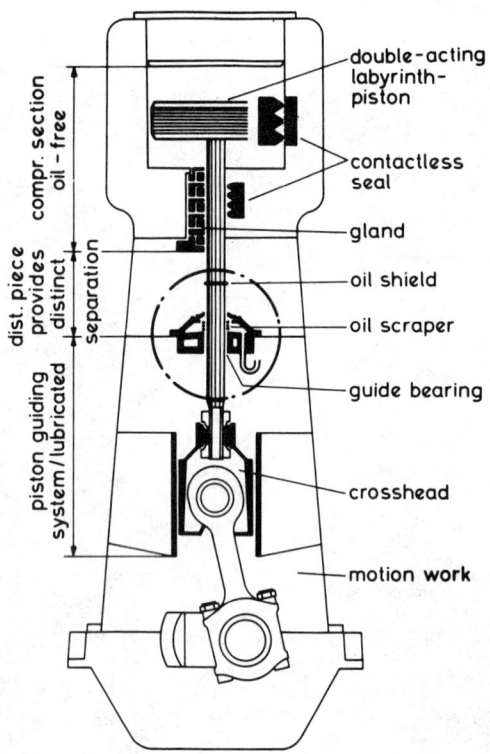

FIG. 2—*Labyrinth-piston compressor design and constructional features. The ring marks the area where the oil detector is placed.*

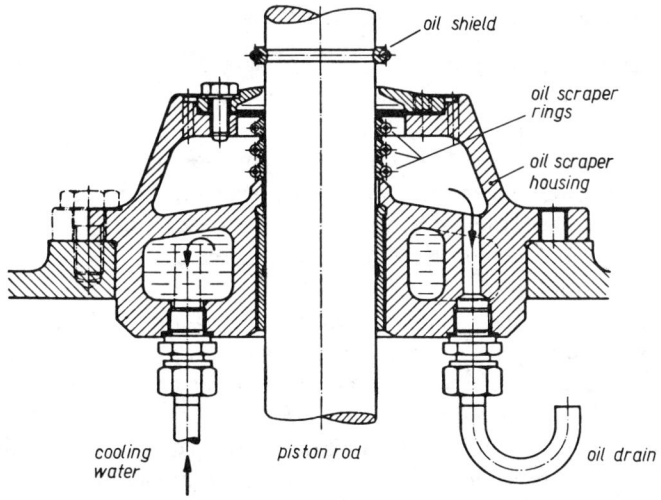

FIG. 3—*Standard piston-rod guide-bearing with oil scraper rings without oil detector.*

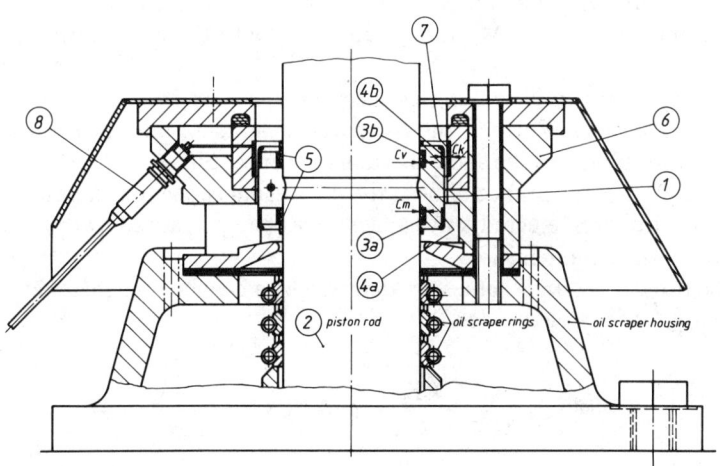

FIG. 4—*Oil detector arrangement.*

any accretion of appreciable quantities of lubricant on the piston rod long
before the first drop of oil could reach the glands.

Measuring Method

The system uses the variation in the capacitance of a capacitor as a result of
absorption of oil by an absorbent element serving as a dielectric.

Detector Design

In place of the conventional oil shield, a detector ring (1) of insulating material (Fig. 4), consisting of two parts, is clamped by two screws on each piston rod (2) of the compressor. To avoid axial slipping of the device on the piston rod, the latter has a slight groove filled by the ring. This ring carries on the inside two annular electrodes (3a and 3b), each of which is connected via a radial conductor with the corresponding annular electrode (4a or 4b) on the outside of the ring. Each of these four electrodes consists of a layer of copper applied by electroplating.

The electrodes (3a and 3b) form, respectively, together with the piston rod, the measuring capacitor (CM) and the comparison capacitor (CV). The annular spaces between the electrodes (3a and 3b) and the piston rod are filled with a ring of lubricant-absorbing material (5). As it approaches the "bottom dead center position (BDC)," the detector ring enters the stationary housing (6) embodying the annular electrode (7). A small radial clearance between the oscillating electrodes (4a/4b) and the stationary electrode (7) is necessary. The air gap between them forms a coupling capacitor (CK), which is equal for both the electrodes (4a and 4b). On the downward movement of the piston rod, the coupling capacitor (CK) is thereby first connected in series with the measuring capacitor (CM) and shortly after with the comparison capacitor (CV).

This system avoids making connections between the stationary and the moving electrodes. Further, the comparison capacitor and the measuring capacitor are subject to the same thermal influence. Since on each stroke of the piston rod the capacitance of both the measuring capacitor and the comparison capacitor are detected, any capacitance changes caused by thermal influences are thus compensated.

The two capacitance values are detected by a capacitance meter at the connection (8) of the electrode (7)

$$CM^* = \frac{CM \cdot CK}{CM + CK} \quad \text{and} \quad CV^* = \frac{CV \cdot CK}{CV + CK}$$

Should the oil scraper rings be no longer intact or have worn, appreciable quantities of lubricant occur on the piston rod. As the absorbent ring (5) of the measuring capacitor (CM) absorbs the lubricating oil, its dielectric constant increases, resulting in a change of the capacitance value (CM*). In comparison with the capacitance values (CV*) of the comparison capacitor (CV), which has remained dry, the connected electronic circuit provides a differential signal corresponding to the amount of lubricant absorbed. This signal change can then be used to shut off the compressor or trigger an alarm or both. The system guarantees a continuous monitoring of the condition on the whole circumference of each piston rod.

According to the number of piston rods in the compressor, there is a corresponding number of identical capacitor arrangements (Fig. 5) connected via a cable (9) to the electronic unit (10) mounted on the compressor. The electronic unit (10) connects the coupling capacitors sequentially to the evaluator unit (11) installed in the control room, which can be up to 100 m away from

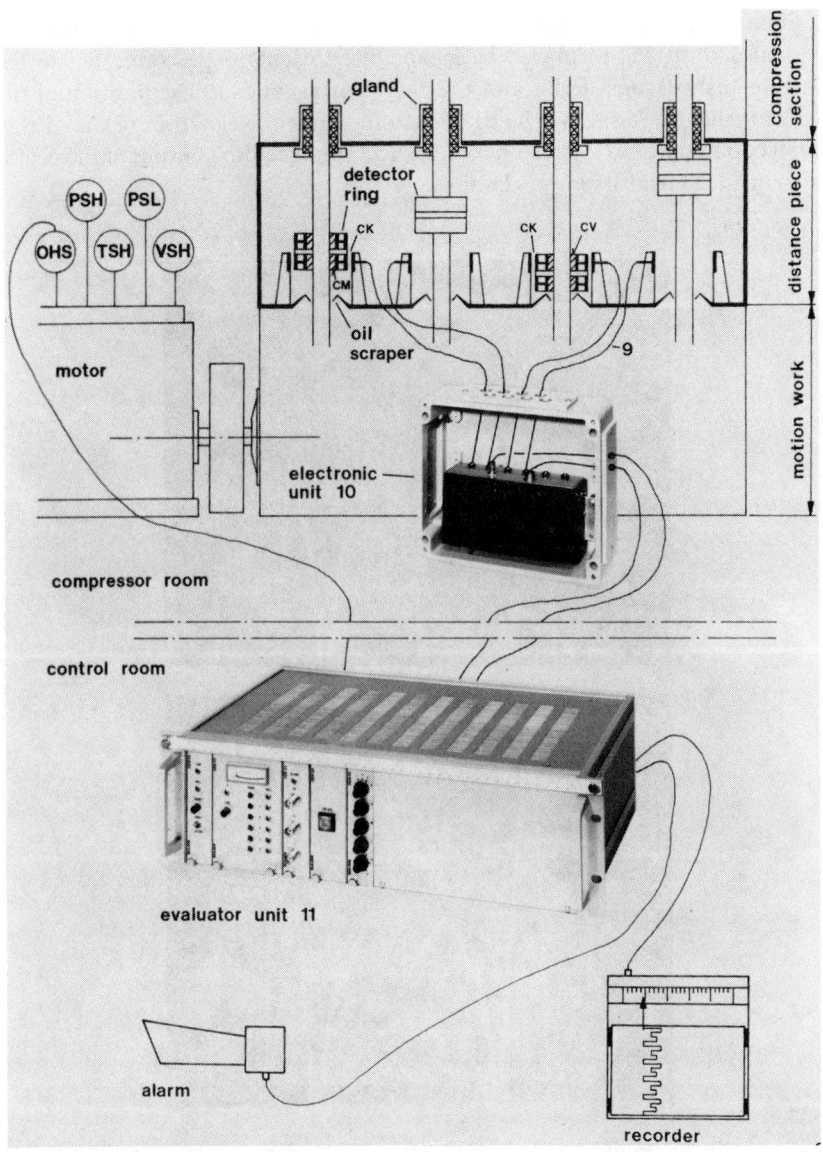

FIG. 5—*Oil detector system.*

the compressor. Control lamps on the evaluator unit indicate which piston rod is currently being monitored. The associated meter indication is automatically held constant for about 15 s to allow the operator to note the measured value. The operator can check the condition of the oil retention system, either by the meter at the evaluator unit or on the paper tape of a connected recorder. In case of an alarm or shutdown of the compressor, the corresponding control lamp shows the piston rod in failure.

Figure 6 shows a graph plotted against time for the voltage signal from the evaluator unit (11) displayed by means of an oscilloscope connected to the unit for test reasons. Point a of the curve corresponds to the position of the detector ring (5) far above the BDC position. Here the electrode (7) has a very small capacitance to the piston rod (2). The voltage at the output of the evaluator unit (11) is at its lowest level.

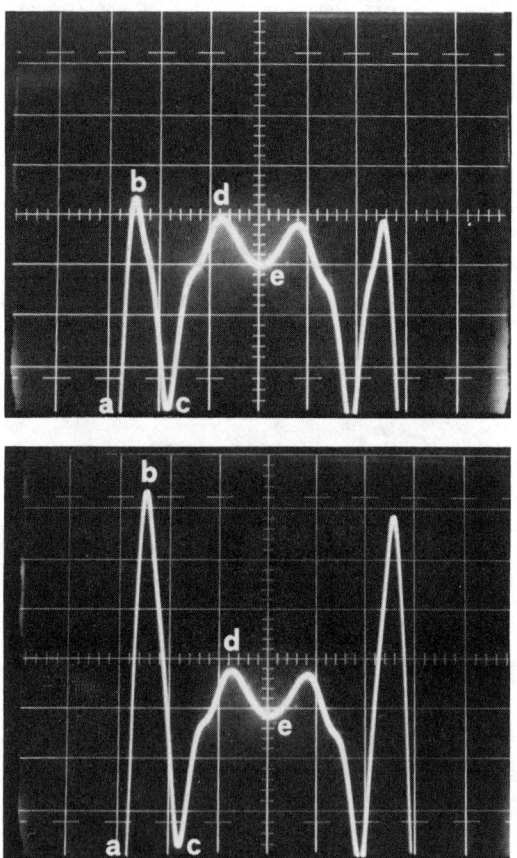

FIG. 6—*Typical oil-detector measuring signal displayed by means of an oscilloscope.* Upper: *absorbent material dry.* Lower: *absorbent material partially oil-saturated.*

On the downward movement of the piston rod, the electrode of the measuring capacitor (CM) first enters the coupling capacitor (CK). The voltage quickly rises positively to the peak, Point b, the height of which depends on the amount of oil absorbed in ring (5).

The base, Point c, of the trough occurs when the electrode (7) is situated between the electrode (4a and 4b).

The next peak, Point d, corresponds to the capacitance of the comparison capacitor (CV). This peak remains at a constant level. The trough, Point e, corresponds with the BDC, and when the piston rod continues to move upwards, the curve is continued in a slightly damped mirror-image arrangement.

A prototype has been in operation for more than two years on a four-throw machine. All tests have been passed very successfully. The first industrially manufactured electronic oil detector systems have now been in operation since Aug. 1984.

The field experience shows that this system is a major step forward in safe, automatic, remote control of oxygen reciprocating compressors.

Barry L. Werley[1]

Quantitative Low-Tech Measurement of Fluid Nonvolatile Residue

REFERENCE: Werley, B. L., **"Quantitative Low-Tech Measurement of Fluid Nonvolatile Residue,"** *Flammability and Sensitivity of Materials in Oxygen-Enriched Atmospheres: Second Volume, ASTM STP 910,* M. A. Benning, Ed., American Society for Testing and Materials, Philadelphia, 1986, pp. 220–229.

ABSTRACT: A relatively simple and inexpensive mechanism is proposed for measuring the volume of fluid, nonvolatile residue (NVR) dissolved in dense solvents. The mechanism involves supporting the solvent/NVR with surface tension in an inverted transparent capillary tube and evaporating the solvent fraction to leave a continuous measurable column of NVR. The mechanism has been briefly tested on several hydrocarbon oils and has exhibited potential sensitivity over a range of at least 0.5 to 50 μL. The mechanism may be applicable to field measurement of oils extracted from surface films, porous insulations, and water. The mechanism is proposed as an embryonic basis for development into a potential ASTM standard.

KEY WORDS: safety, oil film tests, cleanliness tests, nonvolatile residue test, contamination test, oxygen compatibility

Various applications require the measurement of fluid nonvolatile residues (NVRs). Frequently, the NVR has been extracted by a dense solvent such as chlorinated or fluorinated solvents. The analysis of oil films in oxygen systems exemplifies an application where NVRs must be measured. In this case, many different analytical methods are in use. Numerous methods are reviewed by Gilbertson and Lowrie[2] and in Compressed Gas Association Pamphlet G-4.1 [*1*]. However, those methods that are simple are typically not quantitative; conversely, methods that are quantitative tend to require expensive equipment and considerable technical skills. Consider the following examples:

- Black (ultraviolet [UV]) light inspection is often used, but it is not quantitative, and many oils do not fluoresce. This can yield error.

[1]Hazards research specialist, Air Products and Chemicals, Inc., P.O., Box 538, Allentown, PA 18105.
[2]Gilbertson, J. and Lowrie, R. E., in this publication, pp. 204–211.

- White light inspection is commonly used, but it is also nonquantitative, and its detection threshold can be unacceptably high in many cases.
- Wipe tests are also frequently used in which a lint-free cloth is rubbed on a surface to collect oil. The cloth is then inspected under white light or black light. The method is more sensitive, in some cases, than the above tests but is still not quantitative and can be interfered with by discolorations unrelated to oil presence.
- Water break tests are seldom used. The method involves spraying water onto a surface. "Beading" of the water is assumed to indicate an oil film is present. The method is not quantitative and is interfered with by such things as nonhorizontal surfaces, nonsmooth surfaces, surfactant films, or water-soluble oils.
- Extraction tests are frequently used in which a solvent is employed to strip an oil from the subject surface. The solvent is then analyzed for oil content, which is then correlated to the surface area that the oil had occupied. The solvent can then be evaporated and the oil residue weighed. Alternatively, the solvent can be analyzed for oil by chromatograph or infrared analyzer. These two approaches are both quantitative; however, the hardware (analytical balances or electronic instruments) is expensive and requires a skilled operator. An alternative is to evaporate the solvent mixture onto a smooth surface, such as a watchglass, but results at Air Products indicate the procedure is inadequate. Another variation is to greatly scale up the amount of solvent that is evaporated (for example, to 1000 mL) using an Imhoff cone, so that the residue can be volumetrically determined; however, this requires extraction of a very large surface area, involves long evaporation times, and results in venting of large quantities of solvent vapor.

Methods based upon the mechanism described herein would also use solvent extraction, but should not require highly skilled operators, expensive hardware, or large solvent samples. In fact, one preliminary procedure can be learned in hours, the equipment costs can be under $200, and the solvent samples can be as small as a few millilitres. The most important shortcoming currently identified is that an analysis requires roughly one or more hours; but future refinements or alternative procedures may be able to reduce this.

Overview of Mechanism

An extract of solvent with dissolved NVR is evaporated in an inverted glass capillary tube. A large, apparently constant, fraction of the NVR is collected into a continuous cylindrical column that may be viewed much as one views the liquid in a thermometer. The length of the column is proportional to the quantity of NVR present. For the nominal 1-mm bore capillary used in preliminary tests, the theoretical column lengths are about 0.5 mm (0.02 in.) for a 0.4-μL NVR volume and about 50 mm (2 in.) for a 40-μL volume. In preliminary tests, approximately half the theoretical column lengths were observed.

The analytical use of an inverted, surface-tension-supported column of solvent/NVR may be a new approach. One potential procedure that was used to test the possibility of collecting NVR into a capillary is presented in this paper. It has worked in preliminary testing with hydrocarbon oil, but the effect of silicone oil or halocarbon oil is unknown. Work to date suggests a useful procedure may be developed in cases where the nature of the NVR is known, and the procedure can be tested and qualified in advance of its use with the known NVR. For example, oils of known properties are added to rock wool during its manufacture. The oil and its specific gravity are known, and a procedure may be either calibrated in advance using bulk oil, or it may be used with a simultaneous comparison standard.

Numerous other implementations of the mechanism may be possible, such as use with micropipets to extend the range of apparent sensitivity. Standardization may also identify uses for the general analysis of extracted NVR contaminants. However, use of the example procedure below or related procedures is discouraged until a consensus standard is created or until the user has thoroughly qualified its suitability for any intended purpose.

Example Procedure

The following method was used to evaluate the ability to collect NVR in a capillary tube.

Apparatus

- Extraction equipment and solvent as required.
- Glass syringe with inert piston (0 to 5 mL suggested).
- Disposable Pasteur pipets (Fisher 13-678-6B), modified to seal the capillary end by brief exposure of the tip to flame (Fig. 1).
- 1,1,1-trichloroethane or methylene chloride solvent (may be used for extraction, as well as analysis).
- Low heat source (for example, warm water or steam bath, heat gun, hair dryer, or incandescent bulb).
- Vacuum source (such as a vacuum pump or water aspirator). This item can be omitted if several days are available for analysis.
- Manifold with vacuum adjustment valve for evacuating pipets.
- Tubes to slip onto the pipets to water-jacket the capillary gage section (short lengths of clear vinyl tubing are acceptable).

Procedure

1. Extract the NVR from the subject system using an appropriate volatile solvent (trichlorotrifluoroethane, 1,1,1-trichloroethane, and methylene chloride are typical). This can be accomplished in numerous well-known ways, including:

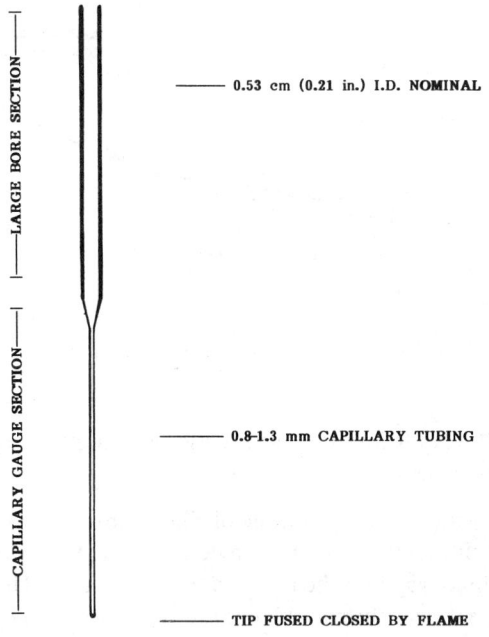

FIG. 1—*Fused Pasteur pipet.*

- flowing solvent over the surface and collecting the runoff,
- using Soxhlet extraction apparatus, and
- placing the solvent and test material (rock wool insulation, contaminated water, and so forth) into a vessel, agitating, and then decanting or otherwise separating the results.

2. If required, concentrate the solvent/NVR solution by evaporation.

3. Draw a sample of solvent/NVR into a clean syringe; a sample size of 1 to 5 mL is suggested.

4. Hold a sealed-capillary pipet slightly inclined from a horizontal position (Fig. 2) while the solvent/NVR is slowly injected. The rate of introduction should be slow enough to allow the solvent to flash, leaving the NVR plated out on the large-bore region of the pipet; rotating the pipet is useful to speed evaporation. The angle of the pipet and the rate of sample introduction should be controlled so that the sample does not flow into the capillary/gage section of the pipet. Inadvertent flow into the capillary should be back-flowed into the large-bore section. To speed solvent evaporation, the pipet can be warmed slightly over a heat source, but overheating *must* be avoided to limit NVR vaporization. The introduction of heat is only intended to provide the latent heat of evaporation to the solvent that may otherwise subcool itself well below ambient temperature.

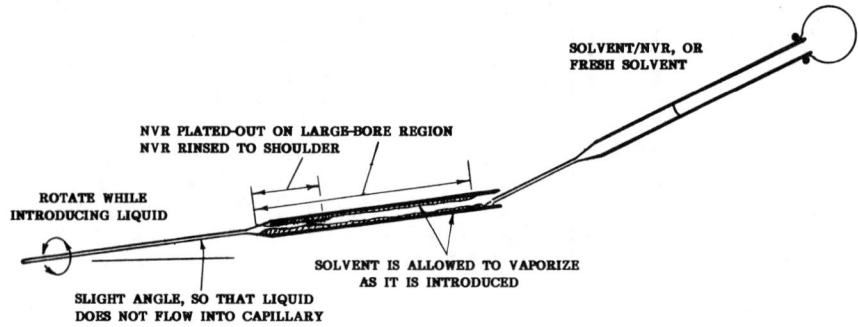

FIG. 2—*Introduction of solvent/NVR or first rinse.*

5. Transport the NVR into the gage section of the pipet with fresh solvent using a two-stage process:

• Using the positions and technique of Fig. 2, introduce fresh chlorinated solvent (1,1,1-trichloroethane and methylene chloride are suitable) dropwise in a carefully limited way into the nearly horizontal pipet. This step rinses the NVR and deposits it near the shoulder of the transition into the capillary gage section; if necessary, the solvent should be vaporized by slightly warming the pipet to prevent evaporative cooling below ambient temperature.

• Then, use a limited quantity of solvent to rinse the NVR into the gage section of the pipet; a clean syringe or second unsealed disposable pipet with a rubber dropping bulb may be used. Carefully flow solvent into the carefully inclined pipet while rotating it, so that the mixture flows into the gage section, and so that the entire NVR film is rinsed (Fig. 3). If the rate is too fast, a meniscus may form and a plug of liquid may block the capillary. Should this happen, drive the plug to the capillary bottom carefully by tapping the large-bore region of the pipet. Use only enough solvent to fill the capillary to the 50 to 80% point.

6. Hold the pipet vertically until all solvent has thoroughly drained into the gage section. There should be no bubble in the capillary fluid column; tap the pipet to clear bubbles.

7. Invert the pipet and insert it into soft rubber tubing connected to the vacuum manifold. Slip a water jacket (such as a length of clear vinyl tubing held in place by friction) onto the pipet and fill the jacket with room-temperature water (Fig. 4).

8. Slowly evacuate the pipet. A vacuum of 6 kPa (24-in. W.C. vacuum) has worked well with methylene chloride, and a vacuum of 7 kPa (28-in. W.C. vacuum) has worked well with 1,1,1-trichloroethane. Allow it to stand for about 1 h. The solvent will vaporize, and the solvent/NVR meniscus will slowly recede up the gage region under the influence of surface tension until a

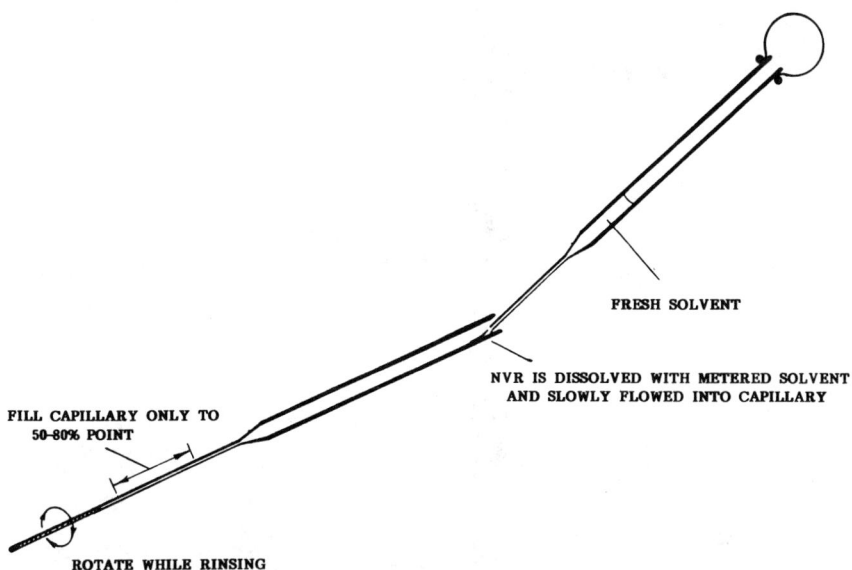

FRESH SOLVENT

NVR IS DISSOLVED WITH METERED SOLVENT
AND SLOWLY FLOWED INTO CAPILLARY

FILL CAPILLARY ONLY TO
50-80% POINT

ROTATE WHILE RINSING

FIG. 3—*Rinsing solvent/NVR into capillary.*

slug of NVR remains at the tip of the capillary. The test is complete when either the column length stops receding or when the column length has decayed to a value less than the cleanliness acceptance criteria. *Inspect the entire pipet for any evidence of meniscus failure.*

9. If optimum accuracy is desired, any small quantity of NVR that has plated-out on the capillary wall as the meniscus receded can be rinsed with a second smaller quantity of solvent, and vacuum evaporation can be repeated.

10. The volume of NVR present is proportional to the length of the column and the column's average cross-sectional area. To achieve optimum accuracy, measure the capillary inside diameter and taper before fusing of the tip.

Example Results

A limited amount of laboratory testing was conducted to confirm that hydrocarbon oil could, indeed, be concentrated within a capillary. Table 1 presents the results of a final demonstration test series.

Before fusing the tips, five Pasteur pipets were selected so that all had the same inside diameter at the capillary outlet, 0.94 mm (0.037 in.); no measurement was made of taper along the capillary length. A mixture of 6.25% by volume of motor oil in methylene chloride solvent was prepared and successively diluted by factors of 2.5, 5, 50, and 100. Samples of 1 mL of each mixture, as well as a blank, were analyzed using the above procedure.

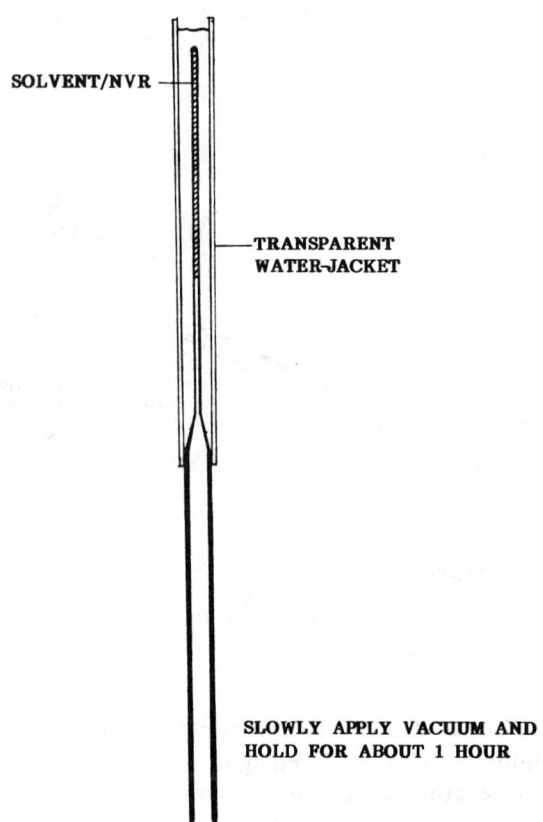

FIG. 4—*Final solvent vacuum evaporation.*

TABLE 1—*Example test results.*

Oil Concentration[a] Volume, %	Dilution Factor[b]	Expected Column Length[c]		Measured Column Length[c]		Measured/ Expected
		cm	in.	cm	in.	
6.25	. . .	9.01	3.55	5.05	1.99	0.56
2.5	2.5	3.60	1.41	1.93	0.76	0.54
1.25	5.0	1.80	0.71	0.99	0.39	0.55
0.125	50.0	0.18	0.071	~0.127	~0.050	~0.7
0.0625	100.0	0.09	0.036	~0.076	~0.030	~0.8
blank[d]	. . .	0	0	0	0	. . .

[a]Concentration in methylene chloride solvent.
[b]Dilution of 6.25% stock.
[c]1-cm^3 sample of oil in methylene chloride with 1,1,1-trichloroethane rinse solvent.
[d]No sample, but 1,1,1-trichloroethane rinsing performed.

Table 1 presents the lengths of oil columns that resulted in comparison to the expected values and the dilution factors that were used. About 55% of the oil appeared to be present. The results exhibit good linearity. Indeed, for an oil of specific gravity of 0.8, these results would be equivalent to oil masses ranging from roughly 0.3 to 30 mg, that assuming the 1-mL oil/solvent sample had been derived by extracting a 0.1-m^2 (1.1-ft^2) surface area, would span the range of many presently used cleanliness criteria for oxygen systems.

Key Elements of Procedure

One key element is vital in rendering this mechanism practical: the effectiveness of vacuum evaporation of an inverted capillary column of chlorinated solvent and NVR.

If a pipet containing solvent/NVR is placed vertically, "without" inversion, and allowed to evaporate with or without vacuum, then three undesirable effects occur:

1. The solvent evaporating from the top of the meniscus creates a progressively more concentrated layer of low density NVR that floats. This layer suppresses the solvent vapor pressure at the surface and greatly inhibits the solvent vaporization (by one or more orders of magnitude).

2. As the meniscus recedes, the floating layer of NVR leaves a very thick film of NVR on the capillary walls that represents significant error in the length of the final column of NVR that forms.

3. Since most solvents have vapors heavier than air, the solvent vapors fill the pipet (unless evacuated to very low pressures). This limits the rate of removal of solvents from the pipet to the rate of diffusion of the heavy vapor out of the pipet.

In contrast, when the pipet is inverted, the solvent/NVR is held in the top of the gage section by surface tension. Any solvent vapors that evolve from the bottom of the meniscus yield an NVR-enriched region of reduced density that immediately migrates upward and tends to mix with the remainder of the solution; layering does not occur. As a result, the solvent concentration at the meniscus is enhanced, yielding a higher vapor pressure and higher evaporation rate. In addition, the inverted arrangement allows the heavy vapors to migrate out of the bottom of the pipet under gravity.

Because the concentration of NVR at an inverted meniscus is greatly reduced, the layer of NVR that plates out on the walls of the capillary is therefore much thinner. Therefore, a greater fraction of the NVR accumulates in the final plug of NVR that forms.

Actually, rapid evolution of the solvent should be possible without pipet inversion, if an effective method could be found for keeping the solution mixed. This would allow simplification of the procedure in a number of ways and would eliminate the possibility of meniscus failure and resultant drainage of the sample. This is an area where improvements are possible. However, one method tested, ultrasonic agitation, has not been effective at mixing the layers. Also, mechanical stirring is possible in principle, but it introduces undesirable additional hardware and complexity. Hence, a practical alternative has not been found.

One other alternative would be to use a solvent, such as ethyl ether, that has a lower density than typical NVR. Theoretically, the NVR should tend to sink to the bottom, causing mixing and allowing vacuum evaporation without inverting the capillary; instead, it appears that an ether film wets the capillary wall a short distance above the meniscus, and the principal evaporation occurs in this film. As a result, a mass of NVR forms above the meniscus that is pushed upward by the evolving ether vapors. Nonetheless, it is possible to transport the NVR to the capillary bottom using several evaporation cycles with progressively smaller quantities of ether in each. However, ether is an undesirable solvent for field use because of its anesthetic properties, its flammability, and its tendency to generate explosive peroxides if stored beyond its shelf-life.

Finally, it should be noted that two fluorocarbon solvents (trichlorofluoromethane and trichlorotrifluoroethane) were also tested in the inverted meniscus configuration but exhibited meniscus failure. These fluorocarbons are dense solvents and are desirable for field use because of their low toxicities; however, they did not perform well in this application. Both fluorocarbons wetted the capillary wall a short distance below the meniscus. The presence of this film allowed a fraction of the solvent/NVR (often the entire quantity) to flow out of the capillary. Although the evacuation appeared to proceed normally, the final result, in these cases, was a visible droplet or ring of NVR that formed near the bottom of the large-bore region of the inverted pipet. Perhaps smaller capillary dimensions or other steps might enable the use of these desirable solvents.

Discussion

Two principal types of quantitative NVR test methods are available: mass or volume measurements. The mechanism described here is a volumetric method. The majority of methods in use measure mass. To obtain mass analysis with the present mechanism requires a knowledge of, or an assumption about, the NVR's specific gravity.

Most existing cleanliness criteria are on a mass per unit surface area basis. For a hydrocarbon NVR (the generally recognized worst case contaminant), the heat release upon combustion is directly proportional to the mass present; hydrocarbons have similar heats of combustion on a mass basis. The risk of damage caused by self-combustion or the ability to promote ignition of other materials in an oxygen system is similarly proportional to NVR mass presence. This argues for a mass NVR analysis method and would require a conservative assumption of NVR specific gravity to be made when volumetric test methods are used.

One can argue, however, that cleanliness criteria should be established on the basis of migration thresholds rather than flammability thresholds. A previous paper [2] noted that NVR migration is likely to occur at surface concentrations well below their flammability thresholds. Indeed, the major risk may be whether the level of NVR present will migrate and accumulate.

Migration thresholds may be more realistically represented by a volumetric criterion. For example, Fote et al [3] argue that for oils on rough surfaces, a migration threshold is crossed when the amount of oil present has filled the micropores in the surface. After the pores are full, surface tension no longer retains the oil in the pores, and migration is greatly facilitated. Clearly, in this case, a volumetric test method would be adequate to appraise cleanliness relative to a volumetric criterion.

Summary

An apparently practical, low-tech mechanism has been demonstrated that allows reasonably quantitative measurement of small quantities of NVR in dense chlorinated solvents. Although there is much work necessary to evaluate the mechanism for practical use, the procedure appears to have merit because of its potentially low cost and simplicity. Optimization efforts and expertise of others may lead to further refinements and a practical standard method. In particular, an understanding of the causes for only fractional recovery of the NVR in the capillary is important. Further, optimization of applied vacuum levels, use of elevated temperatures instead of vacuum, or combinations thereof are potential areas of study.

This embryonic effort is offered to ASTM Committee G-4 on Compatibility and Sensitivity of Materials in Oxygen Enriched Atmospheres for appraisal as a possible analytical approach to determine cleanliness in oxygen systems and related applications. In addition to development of an optimum procedure acceptable to ultimate users, practical field testing and round-robin testing would be required in developing the procedure into an ASTM standard.

References

[1] "Cleaning Equipment for Oxygen Service," Pamphlet G-4.1, Compressed Gas Association, New York, 1977.
[2] Werley, B. L., "Oil Film Hazards in Oxygen Systems," *Flammability and Sensitivity of Materials in Oxygen-Enriched Atmospheres, STP 812*, B. L. Werley, Ed., American Society for Testing and Materials, Philadelphia, 1983, pp. 108–125.
[3] Fote, A. A., Slade, R. A., and Feuerstein, S., "Thermally Induced Migration of Hydrocarbon Oil," *Transactions of the American Society of Mechanical Engineers: Journal of Lubrication Technology*, Vol. 99, No. F-2, April 1977, p. 10.

Author Index

Subject Index